THE BOOK OF PI

C O V E R I N G

$$\pi$$

- THE DEFINITION OF PI
- THE HISTORICAL DEVELOPMENT OF PI
- PI TO 100,000 PLACES

$$\pi$$

Thank you for purchasing this book, we really hope you enjoy it. If you have any queries, suggestions or just want to get in touch, feel free to email me at ben@bclesterbooks.com. All feedback is appreciated on Amazon.

B.C. LESTER BOOKS

THE DEFINITION AND HISTORY OF PI

$$\pi$$

Pi, very simply, is a mathematical constant that describes the ration between the circumference and the diameter of a circle.

$$\pi = \frac{C}{d}$$

The circumference will always be Pi times the amount of the diameter.

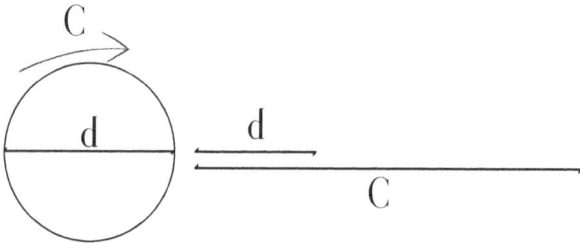

Pi is used in calculating aspects of circles and spheres, including area, surface area and volume, according to the formulae below.

$$A = \pi r^2$$

Where A = area and r = radius

$$V = \frac{4}{3}\pi r^3$$

Where V = volume

$$SA = 4\pi r^2$$

Where SA = Surface Area

Pi is an irrational number, so it cannot be represented as a common fraction, and therefore, also has a recurring decimal that doesn't fall into a repeat pattern. This fact has disallowed the mathematical proof of 'squaring the circle'. The result is a long decimal, that as of 2020 has been calculated to 31 trillion digits. Pi wasn't always this way, and research suggested that Pi was used for practical reasons within Ancient Egypt and Babylonian societies. By the 5th century AD, Chinese Mathematicians estimated Pi to 7 digits. Attempts to find more digits started nearly a millenium later, using infinite series. The first one, was an infinite product (rather than an infinite sum that is more typical for pi calculations) as shown below.

$$\frac{2}{\pi} = \frac{2^2}{2} \times \frac{\left(2+2^{\frac{3}{2}}\right)^2}{2} \times \frac{\left(\left(2+2^{\frac{3}{2}}\right)^2\right)^2}{2} \times \; ...$$

Discovered by a French mathematician in 1593, another infinite product was discovered in Europe in 1655.

$$\frac{\pi}{2} = \left(\frac{2}{1} \times \frac{2}{3}\right) \times \left(\frac{4}{3} \times \frac{4}{5}\right) \times \left(\frac{6}{5} \times \frac{6}{7}\right) \times \; ...$$

After the dawn of calculus, the Gregory–Liebniz discovered this infinite sum:

$$\arctan z = z - \frac{z^3}{3} + \frac{z^5}{5} - \frac{z^7}{7} + \; ...$$

This formula was used by Abraham Sharp with $z = \frac{1}{3}2$ in 1699, the achieve Pi to 71 digits.

This formula convereges too slowly for modern calculations (gets to the answer too slowly), so in 1706, John Machin used a similar formula that converged much faster:

$$\frac{\pi}{4} = 4\arctan\frac{1}{5} - \arctan\frac{1}{239}$$

Machin used this to achieve 100 digits., and due it's fast rate of convergence, this infinite series was used for the subsequent 250 years, where in 1946, Pi was calculated to 620 digits by Daniel Ferguson, the longest approximation without aid from a calculating device. With the dawn of computing, algorithms were used in place of infinite series to take digits of Pi to new records of 1120 in 1949, 10,000 in 1958, 100,000 in 1961 and 1,000,000 in 1973. Algorithms were much faster than infinite series, and iterative algorithms of the 1980s again allowed new records to be broken.

Scientists generally use pi up to a couple hundred digits, and the drive to find more digits is based on record-breaking and also testing supercomputers, and the 31.4 trillion digits we have already found, will likely be an obsolete figure by the time you purchase this book, as this number is growing at an exponential rate.

Pi, as of 2020, has been memorised to 67,890 digits, using the method of loci. If you are here looking to break the record, then confirm this as this number is also ever-changing.

PI TO 100,000 DIGITS

π

3.14159265358979323846264338327950288419716939937510582097494459230781640628620899862803482534211706798214808651328230664709384460955058223172535940812848111745028410270193852110555964462294895493038196442881097566593344612847564823378678316527120190914564856692346034861045432664821339360726024914127372458700660631558817488152092096282925409171536436789259036001133053054882046652138414695194151160943305727036575959195309218611738193261179310511854807446237996274956735188575272489122793818301194912983367336244065664308602139494639522473719070217986094370277053921717629317675238467481846769405132000568127145263560827785771342757789609173637178721468440901224953430146549585371050792279689258923542019956112129021960864034418159813629774771309960518707211349999998372978049951059731732816096318595024459455346908302642522308253344685035261931188171010003137838752886587533208381420617177669147303598253490428755468731159562863882353787593751957781857805321712268066130019278766111959092164201989

93809525720106548586327886593 6 153
38182796823030 19520353018529 68995
7736225994 1389 124972 1775283479 13 1
5155748572424541506959508 29533 116
86172785588907509838 1754637464939
319255060400927701671139009848824
012858361603563707660104710181942
955596 1989467678374494482553797 74
726847 1040475346462080466842590 69
49 129331367702898915210475216 2056
966024058038 150 19351 125338243 0035
58764024749647326391419927 2604269
92279678235478 16360093417216412 19
9245863 15030286 18297455570674 9838
5054945885869269956909272 107975 09
302955321 16534498720275596023648 0
665499 1198818347977535663698074 26
5425278625518184 17574672890977 772
79380008 16470600 16 145249 192173 2 17
21477235014144 19735685481613 61 157
3525521334757418494684385233239 07
394 1433345477624 16862518983569485
562099219222 18427255025425688767 1
7904946016534668049886272327917 86
0857843838279679766814541009538 83
786360950680064225125205 11739298 4
896084128488626945604241965285 022
2106611863067442786220391949450 47
1237 13786960956364371917287467 764
6575739624 13890865832645995813390
4780275900

994657640789512694683983525957098
258226205224894077267194782684826
014769909026401363944374553050682
034962524517493996514314298091906
592509372216964615157098583874105
978859597729754989301617539284681
382686838689427741559918559252459
539594310499725246808459872736446
958486538367362226260991246080512
438843904512441365497627807977156
914359977001296160894416948685558
484063534220722258284886481584560
285060168427394522674676788952521
385225499546667278239864565961163
548862305774564980355936345681743
241125150760694794510965960940252
288797108931456691368672287489405
601015033086179286809208747609178
249385890097149096759852613655497
818931297848216829989487226588048
575640142704775551323796414515237
462343645428584447952658678210511
413547357395231134271661021359695
362314429524849371871101457654035
902799344037420073105785390621983
874478084784896833214457138687519
435064302184531910484810053706146
806749192781911979399520614196634
287544406437451237181921799983910
159195618146751426912397489409071
8649423196

15679452080951465502252316038 8193
0142093762137855956638937 78708 303
906979207734672218256259966150142
150306803844773454920260541466592
52014974428507325186660021324 3408
819071048633173464965145390579626
856100550810665879699816357473638
4052571459102897064140110971 20628
04390397595156771577004203386993
6007230558763176359421873125147 12
05329281918261861258673215791984 1
484882916447060957527069572209175
6711672291098169091528017350 67 127
48583222871835209353965725121 0835
791513698820914442100675103346711
03141267111369908658516398315 0197
0165151168517143765761835 15565088
490998985998238734552833163550764
79185358932261854896321 3293308985
706420467525907091548141 654985946
163718027098199430992448895757128
289059232332609729971208443357326
548938239119325974636673058360414
2813883032038249037589852437 44170
29132765618093773444030707 4692112
01913020330380197621101 1004492932
151608424448596376698389522868478
312355265821314495768572624334418
930396864262434107732269780280731
891544110104468232527162010526522
7211166039

666557309254711055785376346682065
310989652691862056476931257058635
662018558100729360659876486117910
453348850346113657686753249441668
039626579787718556084552965412665
408530614344431858676975145661406
800700237877659134401712749470420
562230538994561314071127000407854
733269939081454664645880797270826
683063432858785698305235808933065
757406795457163775254202114955761
581400250126228594130216471550979
259230990796547376125517656751357
517829666454779174501129961489030
463994713296210734043751895735961
458901938971311179042978285647503
203198691514028708085990480109412
147221317947647772622414254854540
332157185306142288137585043063321
751829798662237172159160771669254
748738986654949450114654062843366
393790039769265672146385306736096
571209180763832716641627488880078
692560290228472104031721186082041
900042296617119637792133757511495
950156604963186294726547364252308
177036751590673502350728354056704
038674351362222477158915049530984
448933309634087807693259939780541
934144737744184263129860809988868
74413260472

15695 162396 586 457302 163 1598 193 195
167353 812974 1677 29478 672422 924654
366800980 676928 23828 06899 64004 824
354037014 163 149658979 409243 237896
907069779 42236 250822 1688957383798
623001593 77647 165122893 57860 15881
617557829 7 352334 460428 15 126272037
343146531977 774 160319906 6554 18763
979293344 1952154 134189948 544 47345
673831624993 4191318 14809 27777 1038
638773431772075456 545322077 709212
0190516609628049092636 0197 598828 1
6133231666365286 193266 86336062735
676303544776280350450 77 235547 105
859548702790814356240145 17 1806246
4362679456 12753181340783 303362542
32783944975382437205 8353 11477 1 199
260638133467768796959703098339 130
771098704085913374641442822772634
659470474587847787201927 71528073 1
767907707 1572134447306057007 33492
436931138350493163 1284042512 19256
517980694 1135280 13 1470 13047816437
885185290928545201 165839341965621
349143415956258658655705526904965
209858033850722426482939728584783
163057777560688876446248246857926
039535277348030480290058760758251
047470916439613626760449256274204
2083208566 1190625454 3372 13 1535958
4506877246

0290 16 1876 6795 2406 1634 25 22577 1954
29 16299 19306 45537 799 1403 73404 3287
52628 8896 3995 8794 757 29 17 464 2635 74
5525 407909 14513 571 113694 1091 19393
25 19107 60208 252026 18798 531 8877 058
4297 259 16778 13 1496990090 192 11697 1
7372 78476 847 2686 08 4900 337 70242 429
1651 30050 05168 3233 6435 0389 5170298
9392 2334 5 172 20 1381 28069 650 11 78440
8745 196 012 12285 99371 623 130 17 11444
8464 0903890 6449 5444 006 19869 075485
160 2632 75052 98349 1874 078 668 08 8183
385 10228 334508 504860 8250 39302 1332
197 155 184 30635 45500 7668 2829 493 04 1
3776 552 79397 51754 6 1395 398 46833936
3830 47461 19966 5385 8 15384 205685 338
6218 672 52334 0283 087 11 232827 892 125
077 12629 46322 9563 989898 935 82 11674
5627 0102 1835 64622 0 134967 15 1881909
73038 11980 04973 407 2396 103 6854 0664
319 3950979 0190699 63955 24530 054 505
8068 550195 67302 292 19 1393 39 1856803
4490 39820 5955 100 226 353 53619 204 199
4745 538 59381 02343 9554 4959 77837790
2374 216 1727 1 11723 6434 354 394 78 2218
1852 86240 851 4006 66044 332 588 56986
7054 3154 70696 5747 4585 5033 23233 42 1
073 01545 9405 16553 7906 8662 73337995
851 1562 5784 3229 882 7372 319898 75714
15957 811196 358330 0594087 3068 12 160
2876496286

744604774649 1599505497 37 425626901
049037781986835938 146574 126804925
648798556 1453723478673303904 68838
34363465537949864 1927056387293 174
872332083760 11230299 11367938 62708
94387993620 16295 15413371424892830
7220 12690 147546684 76535761 6477379
4675200490 75715552781965362 132392
6406 160 136358 15590742202020318727
760527721900556 148425551879253034
35 1398442532234 1576233610 64250639
0497500865627 1095359 194 65897514 13
1034822769306247435363 25691607815
478181 1528436679570611086 15331504
45212747392454494542368288606 1340
84 14863776700961207 151249 14043027
25386074 8236341433462351 89757664
52 164 137679690314950 19 10857598442
39 1986291642 1939949072362346 46844
117394032659 1840443780513338 94525
7423995082 9659 1228508555821572503
1071257012668302402929525220 11872
67675622041542051618416348475 6516
9998 1161410100299607 8386909291603
0288400269 10414079288621507842451
67090870006992821206604 18371 80653
55672525325675328612910424 8776 182
58297651579598470356222629348 6003
415872298053498965022629174878820
273420922224533985626476691490556
2842503912

7577 10284 02799 80663 65825 48892 6488
02545 66101 72967 02664 07655 90429 099
45681 50652 65305 37182 94 12703 369313
78517 86090 40708 667 11496 55834 34347
69338 57817 11386 45587 3678 1230 14587
687 12660 34 89 13909 56200 99393 6 10310
29161 61528 81384 37909 90423 17473 363
94804 57593 1493 14052 97634 75748 1 193
56709 11013 77517 2 100803 15590 248530
90669 20376 7 192 20332 29094 33467 685 1
42214 47737 93937 5 1703 44366 199 10403
37511 17354 719 18550 46449 0263 655 128
16228 82446 25759 16333 03910 72253 837
42182 14088 35086 5739 177 15096 82887 4
78265 69959 9574 49066 17583 44 1375223
97096 83408 00535 59849 1754 1738 18839
99446 97486 76265 5 16582 76584 8358845
31427 75687 90029 095 17028 35297 16344
56212 96404 35231 17600 66510 124 12006
59755 85127 61785 83829 204 1974 844236
08007 19304 57618 93234 92292 79650 198
75187 21272 67507 9812 55470 95890 4556
35792 12210 33346 6974 99235 63025 4947
80249 01141 95212 38281 53091 14079 073
86025 15227 42995 8 1807 247 16259 16685
45133 31239 48049 47079 1191 53267 3430
28244 18604 14263 63954 80004 48002 670
49624 82017 92896 47669 7583 1832 71314
25170 29692 34889 62766 84403 23260 927
52496 03579 96469 25650 49368 18360 900
3238092934

595889706953653494060340216654437
558900456328822505452555640564824
6515187547119621844396582533375438
856909411303150952617937800297412
07665147939425902989695946995657
61218656196733786236256125263208
628692221032748892186543648022967
807057656151446320469279068212073
883778142335628236089632080682224
68012248261177185896381409183903 6
736722208883215137556003727983940
0415297002878307667094447456013 45
564172543709069793961225714298946
71543578468788614445812314593 5719
8492252847160504922124247014 12147
80573455105008019086996033027634 7
87081081754501193071412233908663 9
383395294257869050764310063835198
34389341596131854347546495569781 0
382930971646514384070070736041123
73599843452251610507027056235266 0
1276484830840761183013052793205 42
74628654036036745328651057065874 8
822569815793678976697422057505968
344086973502014102067235850200724
52256326513410559240190274216248 4
39140359989535394590944070469120 9
1409387001264560016237428802109 27
645793106579229552498872758461012
648369998922569596881592056001016
5525637567

8566722796619885782794848855583439
7518744545512965634434803966642055
7982936804352202770984294423253302
2576341807039476994159791594453006
9752148293366555661567873640005366
6564165473217043903521329543529 16
9414599904160875320186837937023488
86894791510716378529023452924 4077
3659495630510074210871426134 97459
5615138498713757047101787957 31042
2969066670214498637464595280 82436
9445789772330048764765241339 07592
0434019634039114732023380715 09522
2010682563427471646024335440 05152
1266932493419673977041595683 75355
5166730273900749729736354964 53328
8869844061196496162773449518 27369
5588220757355176651589855190 98666
5393549481068873206859907540 79234
2402300925900701731960362254 75647
8940647548346647760411463233 90565
1343306844953979070903023460 46147
0961696886885014083470405460 74295
8699138296682468185710318879 06528
7036650832431974404771855678 93482
3089431068287027228097362480 93996
2706074726455399253994428081 13736
9433887294063079261595995462 62462
9707062594845569034711972996 40908
9418059534393251236235508134 94900
4364278527

1383 1591 2568 989 295 1964 27 287 573 946
914 2725 3436 694 153 236 1004 537 304 88 1
985 5170 6594 121 735 2462 5895 487 30 167
6002 9886 5925 786 6285 6 1249 6655 23533
8294 2878 5425 3404 8308 3307 0165 37 228
5635 5915 2534 7844 598 183 134 1 1290 019
9920 5981 3522 05 1173 365 8564 0782 6484
9427 6441 1376 3938 6692 4803 1 18 364 453
698 5891 7544 2647 39988 22846 2 1844 900
8777 6977 63 127 9572 267 2655 5625 96282
542 765 18300 134070 923 3436 5779 160
1280 9317 9401 7185 9859 9933 849 235 495
6400 5709 9558 56 1134 9802 5249 9066 984
2330 1735 0358 0440 81 1685 5265 3 117 099
5708 9942 7328 7092 5848 7894 4364 60050
4108 9226 69 1783 5258 7078 5951 298 344 1
7295 3519 5378 8553 4573 7426 085 902 908
1765 1557 8039 0594 6408 7350 6 123 226 11
2009 373 1080 4854 8526 357 22 825 768 203
4160 5048 4662 7750 4500 3 126 200 800 799
8049 2548 5346 94 1469 7751 6493 2709 504
9346 3938 2432 2271 885 159 7405 4702 148
2897 111 7779 2376 1225 7887 3477 188 196
8254 6298 12 686 858 1705 0740 272 2550 263
3290 4497 627 7894 4236 2 1674 1 19 186 269
4396 5067 15 157 7958 6756 4823 9939 1 760
4260 1763 3870 4549 90 176 1436 4 1204 692
1823 7076 4887 834 196 8968 6 1 18 1558 158
7360 6293 8603 810 17 1215 8552 7266 8300
8238 3404 6564 7588 0405 138 080 1633 638
8742 163 7 14

0643549556 18689641 12282 1407533026
5510042410489678352858829024367 09
04887118190909494533 1442 182876618
10310073547705498 15968077 20094746
96 1343609286 148494 1785017 18077930
6810854690009445899527942439 81392
135055864221964834915126390128038
320010977386806628779239718014613
43244572640097374257007359210 0315
41508936793008 169980536520276007 2
774967458400283624053460372634165
5425902760 183484030681 138185 51059
79705664007509426087885735796037 3
2451414678670368809880609716 42584
975951380693094494015154222219432
913021739125383559 1503 100333032 5 1
117491569691745027 149433 15 1558854
0392216409722910 11290355218157628
2328318234254832611191280 09282525
619020526301639 11477247331485739 1
077758744253876 1 17465786711694147
76421441111263583553871361 0110232
679877564102468240322648346417663
698066378576813492045302240819727
8564719839630878 15432211669122464
159117767322532643356861461865452
2268126887268445968442416107 85401
676814208088502800541436 13 1462308
2102594173756238994207 57 136275167
457318918945628352570441335437585
7534269869

9472547031656661399 1999682628 24727
0641336222 17892390317608542894373
39356 1889 165 125042440400895271983
787386480584726895462438823437517
885201439560057 10481 1949884239060
61369573423 1559079670346 149143447
8863604 103 1823507365027 7859089757
82727313050488939890099239 1350337
325085598265586708924261242947367
0193907727 130706869 170926462 54842
3240748550366080 1360466895 1184009
3668609546325002 14585293095000090
7 15105823626729326453738210493872
4996699339424685516483261 134 14611
0680267446637334375340764 29402668
2973865220935701626384648 52851490
3629320 199 199688285 171839 53669 134
52224447080459239660 28 17 156551565
6661 11359823 11225062890585491 4509
7 157553900243931535 190902 107 11945
73002438801766 15035270862 60253788
179751947806 10 137 150044899 1721002
220 1335013 1060 1639 15415895 7803711
7 79277522597874289 19 179 15522417 18
9585361680594741234 193398420 21874
564925644346239253 1953 135 10331 147
639491 1995072858430658 36 193536932
96992898379 1494 19394060857 2486396
883690326556436421664425760791471
0869984315733749648835292 76932822
0762947282

38 1537 4099 6154 5598 79 8259 89 1093 717
1262 1828 30 258 481 1238 90 1196 822 1429
4576 675 807 1865 380 650 6487 026 1338 92
8229 9497 257 4530 3328 389 638 1843 9447
7077 9402 2843 5988 34 1003 5838 5423 897
3542 4395 647 5556 840 952 2484 4554 1392
394 1000 1620 769 363 6846 7764 130 17819
6593 7997 1557 4685 4194 633 4893 748 439
1297 4239 1433 6593 604 1003 5234 377 706
5888 6778 11 3949 86 1647 8747 1407 93 263
8587 3862 4732 8896 4564 3598 774 667 638
4794 665 0407 4 11 1825 6583 788 7845 4858
1489 6296 1273 9984 1344 2726 086 061 872
4554 5236 0643 1537 10 1127 4680 9778 704
4640 9475 8280 3487 6975 894 8328 241 239
2929 6058 294 861 919 6670 918 958 089 833
2012 1031 8430 340 1284 951 16 2035 34 280
144 1276 1728 5830 2435 598 300 3204 2024
5120 7287 253 558 1195 840 149 180 969 253
3950 7577 8400 0674 655 260 3144 6167 050
8276 8277 2223 534 191 102 634 163 157 147
4061 2385 0425 8459 884 1990 761 1287 258
059 1139 3568 960 1431 668 283 176 323 567
3254 1707 342 081 7332 2304 6298 7992 804
9085 1409 4790 3688 7868 7894 930 546 955
7030 7261 9009 5020 764 334 9335 9106 024
5450 8645 3628 935 4568 6295 853 13 15 337
1838 6826 56 1786 2273 637 1697 5774 1830
2398 600 659 1481 616 4049 449 650 117 321
3138 9574 7062 0884 7480 2365 37 1031 150
8984 2799 242

799275442685327797431139514357417
221975979935968525228574526379628
961269157235798662057340837576687
388426640599099350500081337543245
463596750484423528487470144354541
957625847356421619813407346854111
766883118654489377697956651727966
232671481033864391375186594673002
443450054499539974237232871249483
470604406347160632583064982979551
010954183623503030945309733583446
283947630477564501500850757894954
893139394489922161255255977 0143685
894358587752637962559708167764380
012543650237141278346792610199558
52247172201777237004178084 1942394
872540680155603599839054898572354
674564239058585021671903139526294
455439131663134530893906204678438
778505423939052473136201294769187
497519101147231528932677253391814
660730008902776896311481090220972
452075916729700785058071718638105
496797310016787085069420709223290
807038326345345203802786099055690
013413718236837099194951648960075
504934126787643674638490206396401
976668559233565463913836318574569
814719621084108096188460545603903
845534372914144651347494078488442
3772175154

334 260 306 698 83 17 68 33 100 1 133 108 690
421 939 031 080 143 784 334 151 370 924 353
013 677 631 084 9 135 16 156 422 698 475 074
303 297 167 469 640 666 53 1527 035 325 467
112 667 522 460 55 119 958 1831 963 763 707
617 99 191 920 357 958 2007 595 605 302 346
267 579 439 363 074 630 569 010 80 114 942
714 1009 391 369 138 107 258 1378 1357 894
0055 995 00 1835 425 1184 1721 360 557 275
221 035 268 037 357 265 279 224 1737 36 057
511 278 872 181 908 449 006 1780 13 889 710
770 822 931 002 797 665 935 838 758 909 395
688 148 560 263 224 393 726 562 472 776 037
890 814 458 837 8550 1970 2843 779 362 407
825 052 704 875 816 470 324 581 290 878 395
232 453 237 896 029 841 669 225 489 649 715
606 981 192 186 584 926 770 403 956 481 278
102 1799 132 174 163 058 10 554 598 801 300
484 562 997 651 12 124 1536 374 515 005 635
070 1278 159 267 1424 1342 10 330 1566 165
356 024 733 807 843 028 655 257 222 753 049
998 837 015 348 793 008 062 601 809 623 815
161 366 903 341 11 138 653 85 109 19 367 393
835 229 345 888 322 550 887 064 507 539 473
952 043 968 079 067 086 806 445 096 986 548
801 682 874 343 786 126 453 8 1583 428 0753
061 845 485 903 7982 1799 459 968 1154 419
742 536 344 399 602 902 5100 1588 8272 164
745 006 820 704 1937 61 584 547 123 183 460
072 629 339 550 548 239 557 137 256 840 232
2682 130 124

7679452264482091023564775272308208
810635188991526928891084555711266
0396503439789627825001611015323511
6051965590421184494990778999200073
29476905868577878720982901352956
1397888486050978608595701773129811
5531495168146717695976099421003611
83559138777817698458758104466283
98806006162298486169353373865787
359833616133841338536842119789389
001852956919678045544828584837011
70967212535338758621582310133103
77668272115726949518179589754693
9264219791552338576623167627547
03546994148929041301863861194391
628388705436777432242768091323654
4948536676800000106526248547305 58
6159899914017076983854831887501 42
9389089950685453076511680333732 22
6517566220752695179144225280816511
716677667279303548515420402381746
08923283917032754257508677655 11785
9395002793389592057668278967764 45
31840404185540104351348389531201 3
263783692835808271937831265496174
599705674507183320650345566440344
904536275600112501843356073612227
65949278393706478426456763388 1880
756561216896050416113903906396016
202215368494109260538768871483798
9559999112

099 16464644 1 19 185682770045742 4343
402 1672276 445589330 12778 158686952
50694993646 10 17568 5060 167 145 35 43 1
58 1480 1054588605645501 33203758645
48584032402987 17093480910556 21 167
15468484778039447569798042 63 18099
175642280987 399876697323769573 70 1
58080682290459 92 12366 16890 2596273
043067931653 114940 176473769387 35 1
409336 183321614280214976 3399 18983
548487562529875242387307755955595
546519639440 1821840998412489 82623
67377 1467 22606 163364329640633 5728
1070788758 164043814850 1884 1143188
598827694490 1193212968 27 158884 133
8694346828590066640806314077 75772
57056307294004929403024204984 1656
5479736705485580445865720227 63784
0466823379852827 1057843197535 4179
50 113472736257740802 13476826 04502
285 1579795797647467022840999 56 160
1569 10890384582450267926594205550
3958792298 1852648007068376504 1836
56209455543461 35 13415257006597488
19 163413595567 19649654032 18727 160
264859304903978748958906612725079
482827693895352 1753621850 79629778
5146 1884327 19223 2238 10 15874445052
8665238022532843 89 13752738458 9238
4422535472653098 17 1578447834 21582
2327020690

287232330053862163479885094695472
004795231120150432932266282727632
177908840087861480221475376578105
819702226309717495072127248479478
169572961423658595782090830733233
560348465318730293026659645013718
3754288975557971449924654038681799
213893469244741985097334626793321
0726868707680626399119361965044099
542167627840914669856925715074315
740793805323925239477557441591845
821562518192155233709607483329234
921034514626437449805596103307994
145347784574699992128599999399612
28161521931488876938802228108300 1
986016549416542616968586788372609
587745676182507275992950893180521
872924610867639958916145855058397
274209809097817293239301067663868
240401113040247007350857828724627
134946368531815469690466968693925
472519413992914652423857762550047
485295476814795467007050347999588
867695016124972282040303995463278
830695976249361510102436555352230
690612949388599015734661023712235
478911292547696176005047974928060
721268039226911027772261025441492
215765045081206771735712027180242
968106203776578837166909109418074
4878140490

755 178 203 856 539 099 10 477 594 14 132 15
432 844 062 503 018 027 57 169 650 820 964 2
734 84 146 957 263 978 842 560 084 53 12 140
659 358 090 4 127 113 592 004 197 598 513 62
547 961 606 322 887 361 813 673 732 445 060
792 441 176 399 759 746 193 835 845 749 159
880 976 674 470 930 065 463 424 234 606 342
374 746 660 804 317 0 126 005 205 592 849 36
959 414 340 8 146 852 98 150 539 47 1 789 004
518 357 551 541 252 235 905 906 872 648 786
357 525 419 1 128 887 37 17 663 748 602 766
063 496 035 367 947 026 923 229 7 18 683 277
173 932 361 920 077 7 452 2 1262 475 186 983
349 515 101 986 426 988 784 7 17 193 966 497
690 708 252 174 233 656 627 259 284 406 204
302 141 137 199 227 852 699 846 988 477 023
238 238 400 556 555 178 890 876 613 601 304
770 984 386 1 168 705 23 10 553 149 1625 172
837 327 286 760 072 481 729 876 375 698 163
354 150 746 088 386 636 406 934 704 372 066
886 512 756 882 66 1497 307 886 570 156 850
169 186 474 885 41 679 154 596 507 234 2877
306 998 537 139 043 002 665 307 839 877 638
503 238 182 155 355 973 235 306 860 430 106
757 608 389 086 270 498 418 885 951 380 910
304 235 957 82 495 1439 885 901 131 858 358
406 674 723 702 97 1497 850 841 458 530 857
813 391 562 707 603 563 907 639 473 114 554
958 322 669 457 024 941 398 316 343 323 789
759 556 808 568 362 972 538 679 132 750 555
4 252 449 194

3589128405045226953812179131911451
350099384631177401797151228378546
011603595540286440590249646693070
776905548102885020808580087811577
381719174177601733073855475800605
601433774329901272867725304318251
975791679296996504146070664571258
883469797964293162296552016879730
003564630457930884032748077181155
533090988702550520768046303460865
816539487695196004408482065967379
473168086415645650530049881616490
578831154345485052660069823093157
776500378070466126470602145750579
327096204782561524714591896522360
839664562410519551052235723973951
288181640597859142791481654263289
200428160913693777372229998332708
208296995573772737566761552711392
258805520189887620114168005468736
558063347160373429170390798639652
296131280178267971728982293607028
806908776866059325274637840539769
184808204102194471971386925608416
245112398062011318454124478205011
079876071715568315407886543904121
087303240201068534194723047666672
174986986854707678120512473679247
919315085644477537985379973223445
612278584329684664751333657369238
7201464723

6794278700425032555899268843495928761240075587569464137056251400117971331662071537154360068764773186755871487839890810742953094106059694431584775397009439883949144323536685392099468796450665339857388878661476294434140104988899316005120767810358861166020296119363968213496075011164983278563531614516845769568710900299976984126326650234771672865737857908574664607722834154031144152941880478254387617707904300015669867767957609099669360755949651527363498118964130433116662774712338817406037317439705406703109676765748695358789670031925866259410510533584384656023391796749267844763708474978333655579007384191473198862713525954625181604342253729962863267496824058060296421146386436864224724887283434170441573482481833301640566959668866769563491416328426414974533349999480002669987588815935073578151958899005395120853510357261373640343675347141048360175464883004078464167452167371904831096767113443494819262681110739948250607394950735031690197318521195526356325843390998224986240670310768318446607291248747

403 16 179 69 94 1139 73 877 658 99 868 554 1
703 188 477 886 759 290 260 700 432 126 666 1
79 192 235 209 382 278 788 80 988 633 599 11
608 192 353 555 704 646 349 113 208 591 897
96 13 279 13 197 564 909 760 001 399 623 444
55 350 143 464 268 604 644 958 624 769 0943
470 482 932 94 1404 11 1465 409 239 883 444
35 159 133 20 10 773 944 11 184 074 1076 849
8106 634 724 10 48 239 358 27 40 194 493 566
5 161 088 463 1256 785 297 7697 346 843 030
6 1462 4 180 358 529 331 597 345 830 384 554
103 370 109 167 677 637 42 762 102 1370 135
485 445 0926 307 190 1147 3184 857 492 33 1
8 1672 0721 372 793 556 795 284 439 254 815
609 13728 128 4063 330 393 735 624 200 160
45 664 5574 145 88 166 052 1666 087 387 480
472 433 912 1295 587 776 390 69 690 370 788
2852 775 389 405 246 0758 496 23 157 43 69 1
7 1131 7613 478 388 27 194 1686 066 257 210
3685 132 156 647 800 147 67 523 103 935 786
0689 611 125 996 028 18 393 095 487 090 590
7386 135 19 145 918 195 1029 732 78 755 710
497 290 1148 717 1897 1800 4696 169 77 700
179 139 196 137 914 171 627 070 189 584 692
1434 3696 762 929 27 459 1099 400 600 849 835
684 252 019 1559 370 370 101 10497 473 394
938 778 859 894 17 4330 317 853 487 076 032
2 198 297 0579 751 19 144 05 109 942 358 830
345 463 534 923 498 2688 362 40 433 272 674
1554 030 16 195 0568 0654 1809 394 099 820
2060 999 414

0216890900708213307230896621197 75
53066591881411915778362729274 6156
18571037217247100952142369648 3086
410259288745799932237495519122195
19034244523075351338068568073 5446
49951272031744871954039761073 0806
02699062580760202927314552520 7807
99141842906388443734996814582 7337
20726639176702011830046481900 0241
30835088465841521489912761065 1374
15394356572113903285749187690 944 1
370209051703148777346165287984823
53382972601361109845148418238 0812
05409961252745808810994869722 1612
852489742555551607637167505489617
30168096138038119143611439921 0638
00508321409876045993093248510 2516
82944672606661381517457125597 5495
35802399831469822036133808284 9935
67055755247129027453977621404 9318
20146580080215665360677655087 8380
43041343105918046068008345911 3664
08348874080057412725867047922 5831
91274157390809143831384564241 5094
08491339180968402511639919368 5322
55573389669537490266209232613 1885
58915808324555719484538756287 8612
88590041060060737465014026278 2402
73469625282171749415823317492 3968
35301361786536737606421667781 3773
9951006589

52887742766263684183068019080460984980946976366733566228291513235278880615776827815958866918023894033307644191240341202231636857786035727694154177882643523813190502808701857504704631293335375728538660588890458311145077394293520199432197117164223500564404297989208159430716701985746927384865383343614579463417592257389858800169801475742054299580124295810545651083104629728293758416162532562516572498078492099897990620035936509934721582965174135798491047111660791587436986541222348341887722929446335178653856731962559852026072947674072616767145573649812105677716893484917660771705277187601199908144113058645577910525684304811440261938402322470939249802933550731845890355397133088446174107959162511714864874468611247605428673436709046678468670274091881014249711149657817724279347070216688295610877794405048437528443375108828264771978540006509704033021862556147332117771174413350281608840351781452541964320309576018694649088681545285621346988355444560249556668436602922195124830910605377201980218310103

704 17838 66544 7 18 1260 3971 906 884 623
7085 7518 0800 35327 047 18 56594 994 76 1
2424 81 10999 2886 79 158 9690 4956 39 476
2460 8424 06593 0948 6215 0769 0314 9870
20673 5338 4834 9550 8363 660 178 487 710
60809 804 269 247 1324 1000 946 40 143 736
0326 5645 18 456 679 245 6669 55 100 15022
98330 7984 960 799 4988 249 70 617 236 744
936 122 622 296 17 908 143 114 1466 094 123
4 1593 5930 9585 407 9 1390 872 083 3227 335
49572 080 757 165 17 18 765 994 498 569 379
5623 8755 516 17 575 4380 917 805 2802 946
42004 472 1539 628 0746 360 21 132 942 559
16002 570 735 628 12638 733 10600 589 106
52457 0802 4 47 4937 543 184 14 940 148 21 1
9996 276 453 1068 0066 31 1 838 2376 16 396
6318 093 144 467 129 86 1552 759 820 145 14
10275 600 689 297 502 4630 401 735 148 919
45763 607 893 528 555 0053 17 33 14 1645 705
0499 644 389 0936 3084 387 448 478 396 168
4051 845 273 2884 0323 4520 24 705 685 164
657 164 771 393 237 755 17 294 795 126 1 323
9822 960 239 45 485 797 545 865 174 587 877
13318 1387 529 598 094 12 17 422 730 03 522
9650 808 917 770 506 8259 2488 223 221 549
3804 837 145 47 816 472 139 768 209 633 205
0830 564 792 048 2085 920 475 4998 573 203
8887 639 160 1995 240 9 189 389 455 767 687
49730 8569 559 580 1065 9526 503 036 266 1
5975 066 222 508 4067 428 898 2659 075 106
3756 356 996

82 115 109 49 66 69 74 45 80 54 72 88 69 36 3 102
03 67 82 32 50 18 23 23 70 84 59 79 01 11 54 847
20 876 182 124 778 13 26 66 330 41 20 76 21 658
73 129 708 11 230 758 159 82 124 86 398 072 1
24 078 688 78 11 450 1655 825 13 617 890 307
086 087 019 897 588 980 745 664 3955 15 74 1
536 319 319 198 107 057 533 663 373 803 827
215 279 884 935 039 74 800 158 905 1942 087
971 130 805 123 393 322 190 346 624 991 716
915 094 854 140 187 106 035 460 379 464 337
900 589 095 772 118 080 446 574 396 280 618
67 178 61 017 1567 409 676 620 802 957 6657
705 129 120 990 794 430 463 289 294 730 615
951 043 090 222 143 937 1849 560 634 056 18
934 251 305 726 829 146 578 329 334 052 463
502 892 917 547 087 256 484 260 034 962 961
165 413 823 007 73 133 272 983 050 0160 256
724 014 185 15 204 189 070 1154 288 579 920
81 219 844 93 156 990 591 820 1 18 1973 350
012 618 772 803 68 1248 195 877 070 020 753
240 636 1259 3134 385 955 425 477 8196 114
293 516 356 122 349 666 152 261 473 539 967
405 158 499 860 355 295 332 924 575 238 881
013 620 234 762 466 905 58 1643 896 786 309
762 736 550 472 434 864 307 121 849 437 348
530 060 638 764 456 627 218 666 170 123 812
77 1562 1379 746 1498 6 132 874 411 771 455
244 470 899 714 452 288 566 294 244 023 018
479 120 547 849 857 452 163 469 644 897 389
206 240 194 351 83 1008 828 348 024 924 908
5403077863

8751 1659 1 1302 873 958 787 098 100 772 718
2718 745 290 1397 28 366 14 842 142 871 705
5317 965 430 765 045 343 246 00 536 361 472
6181 809 699 769 334 862 640 774 351 999 28
6863 238 350 887 566 835 950 972 655 748 15
4319 401 955 768 504 372 480 010 204 13 749
8318 722 596 773 871 549 583 997 184 44 907
2791 419 658 459 300 839 426 370 208 756 35
3982 169 620 553 248 032 122 674 989 11 402
6785 285 996 734 052 4203 1091 797 899 905
7188 2194 939 132 075 343 17 07 980 023 736
5909 853 755 202 389 11 643 467 185 582 906
8537 1189 795 262 623 449 248 339 249 6342
4497 1465 684 659 12 489 185 566 629 589 329
9090 352 392 333 336 47 435 203 707 701 010
8438 800 329 075 983 42 170 185 542 283 86 1
6172 104 176 030 11 645 918 780 539 367 447
4720 599 850 235 828 918 336 929 223 37 323
9994 804 371 084 1965 947 316 265 482 5748
0994 825 0999 1833 006 976 569 367 159 689
3644 933 488 647 442 135 008 407 00 660 883
5972 350 395 323 40 179 58 255 70 360 16 936
9909 886 711 32 1097 988 970 705 17 280 755
8551 912 699 306 730 992 507 040 70 245 568
5077 867 906 94 766 126 29 808 225 1633 136
3995 211 709 845 280 926 303 759 22 426 742
5755 998 928 927 837 047 444 52 189 363 203
4894 1552 104 459 726 188 380 030 06 77 617
9313 813 99 1620 580 627 016 51 024 458 869
2476 492 468 919 246 12 1253 10 275 731 390
8404 700 071

4356 1362 3 1699 237 1694 84 81 3255 542 009
1453 04 1037 1354 53 2966 206 392 1054 798
2439 2 1251 7 2540 1323 1490 27 405 858 920
632 1758 9494 345 4890 6846 3993 1375 709
103 4633 27 14 153 1622 3280 5522 9729 795
380 1880 1628 590 735 7 2955 41 627 886 764
982 7418 61 642 1878 9885 74 107 1649 069 1
9 1851 1628 15 285 486 794 17 363 89066 538
857 6422 9 1583 4250 0673 6124 5384 91 606
74 1373 40 173 572 77 9956 34 10 433 268 835
69507 8 1493 1378 007 3623 541 800 706 19 1
8026 732 855 11 9194 267 609 122 10359 874
6924 1 1728 3749 3 126 1633 9500 1239 5992
4050 8454 37 5698 5079 57 0462 2266 46 190
00 1035 00490 183 034 1535 4584 2833 7643
781 1 1988 5563 1877 779 25372 01 1667 185
3954 1835 984 4383 052 037 628 194 407 615
94 1068 207 1697 0302 285 1522 5057 31 260
9304 6898 42 3433 1527 32 13 136 12 165 828
0807 521 263 1547 730 604 4237 747 535 059
52287 1744 0266 6389 14 88 171 73086 436 1
1138 9069 420 279 088 143 1 1944 8799 4 17 1
5404 2 1034 1219 0847 0940 802 540 239 329
4294 5493 87 86 402 305 12927 1 190 975 135
36000 92 197 1105 4120 9668 31 115 163 287
0542 3028 47 0073 120 6580 326 264 171 16 1
6595 76 1327 235 1566 6625 366 727 189 985
34 199 8952 3688 4830 999 302 7574 199 164
6384 142 7077 9887 0887 42 2927 7053 8912
27 1724 863 22028 8984 25 12528 72 178 260
3050 099 45 1

0824783572905691988555467886079 46
2805371227042466543192145281760 74
1482403827835829719301017888345 67
4167811398954750448339314689630 76
3396657226727043393216745421824 55
7062524797219978668542798977992 33
9579057581890622525473582205236 42
4850783407110144980478726691990 18
6438822932305382318559732869780 92
2253529591017341407334884761005 56
4018242392192695062083183814546 98
3923664613639891012102177095976 70
4908305081854704194664371312299 69
2358895384930136356576186106062 22
8705599423371631021278457446463 98
9738188566746260879482018647487 67
2727222062676465338099801966883 68
0994159075776852639865146253336 31
2450536402610569605513183813174 26
1184420189088853196356986962795 03
6738424313011331753305329802016 68
8817481342988681585577810343231 75
3064784983210629718425184385534 42
7620128234570716988530518326179 64
1178579608888150329602290705614 47
6220915094739035946646916235396 80
9201394578175891088931992112260 07
3928149169481615273842736264298 09
8234063200244024495894456129167 04
9508235812487391799648641133480 32
4757775219

708932772262349486015046652681439
877051615317026696929704928316285
504212898146706195331970269507214
378230476875280287354126166391708
245925170010714180854800636923259
462019002278087409859771921805158
532147392653251559035410209284665
925299914353791825314545290598415
817637058927906909896911164381187
809435371521332261443625314490127
454772695739393481546916311624928
873574718824071503995009446731954
316193855485207665738825139639163
576723151005556037263394867208207
808653734942440115799667507360711
159351331959197120948964717553024
531364770942094635696982226673775
209945168450643623824211853534887
989395673187806606107885440005508
276570305587448541805778891719207
881423351138662929667179643468760
077047999537883387870348718021842
437342112273940255717690819603092
018240188427057046092622564178375
265263358324240661253311529423457
965569502506810018310900411245379
015332966156970522379210325706937
051090830789479999004999395322153
622748476603613677697978567386584
670936679588583788795625946464891
3766521995

8828693380 18360 1 19323685 78558558 1
9555604215625088365020332202 45 137
621582046181067051953306530606065
010548871672453779428313388716313
9559690583208341689847606560 71 183
471362181232462272588419902861420
87284956879639325464285343075301 1
052857138296437099903569488852851
9040295604734613113826387889755 17
885604249987483163828040468486 189
38189590542039889872650697620 2019
9554841265000539442820393012748 16
3815853039643992547020167275 93285
7436666164411096256633730540 92195
196751483287348089574777752783442
210910731113518280460363471981856
555729571447476825528578633493428
584231187494400032296906977583 159
038580393535213588600796003420975
4739229673331064939560 18122378 128
545843176055617338611267347807458
50676063048229409653041 1183066710
818930311088717281675195796753471
885372293096 16 143204006381322 4658
41111157758358581135018569047 8153
689381377184728147519983505047812
97718599084707621974605887 4232569
9582889253504193795826061621 18423
687685114183160683158679946016520
577405294230536017803133572632670
5479033840

125730591233960188013782542192709
476733719198728738524805742124892
118347087662966720727232565056512
933312605950577772754247124164831
283298207236175057467387012820957
554430596839555568686118839713552
208445285264008125202766555767749
596962661260456524568408613923826
576858338469849977872670655519185
446869846947849573462260629421962
455708537127277652309895545019303
773216664918257815467729200521266
714346320963789185232321501897612
603437368406719419303774688099929
687758244104787812326625318184596
045385354383911449677531286426092
521153767325886672260404252349108
702695809964759580579466397341906
401003636190404203311357933654242
630356145700901124480089002080147
805660371015412232889146572239314
507607167064355682743774396578906
797268743847307634645167756210309
860409271709095128086309029738504
452718289274968921210667008164858
339553773591913695015316201890888
748421079870689911480466927065094
076204650277252865072890532854856
143316081269300569378541786109696
920253886503457718317668688592368
1488475276

4984688219497397297077371871884 00
4143231276365048145311228509900 20
7424092558592529261030210673681 54
3470152523487863516439762358604 19
1941296976904052648323470009911 54
2426012734380220893310966686367898
6949779940012601642276092608234 93
0411806438291383473546797253992 62
3387915829984864592717340592256 20
7491053085315371829116816372193 95
1887009577881815868504645076993 43
9409874335144316263330317247747486
8979182092394808331439708406730 84
0795893581089665647758599055637 69
5252326536144247802308268118310 37
7358870892406130313364773710116 28
2146146616794040905186152603600 92
5219472188909181073358719641421 44
4786548995285823439470500798303 88
5388608310357193060027711945580 21
9119428999227223534587075662469 26
1776631788551443502182870266856 10
6650035310502163182060176092179 84
6849368631612937279518730789726 37
3537171502563787335797718081848 78
4588665043358243770041477104149 34
9274384575871071597315594394264 12
5702709651251081155482479394035 97
6811881172824721582501094960966 25
3933953809221955919181885526780 62
1499231727

63163218333989696938075616855911 7529
98450132067129392404 14459386239 88
09381240452191484831646210 1473891
82510109096773869066404158973610 4
764365000680771056567 184862814963
7111883219244566394581449 14861655
0049567698269030891111856879869294
705135248160917432430153836847072
9289898284602223730145265567 98986
27767968091469798378268 7643115988
321090437156112997665215396354644
2086919756737000573876497843 76862
876817924974694384274652563163230
0555130417422734 164645512781 27845
77772457520386543754282825671 4128
858345444351325620544642410110379
55464190581168623059644769587 0540
721419852121067343324107567675758
184569906930460475227701670056845
4396923404171108988899341635058 51
57887353430815520811772071 880379 1
0404698306957868547393765643 36319
79786803671873079693924236321 4484
5035477631567025539006542311 79201
5346497792906624150832885839 52905
42637687668968805033317227800 1858
8506973623240389470047189 76193473
443084374437599250341788079 22358
5913424581314404984770173236 16947
197657153531977549971627856631190
4691260918

259 1249890367654 176979903623755 28
65263757337635 2696934435440047 306
719886890 19681474287677908 6697968
85225016369498567302 1752313252926
537589641517 147955953878427849986
64563028788319620998304945 1987439
63690706827626574858 10439112233261
87940599415540632701319898957 0376
11053236062986748037791537675 1158
30432084987209202809297526498 1256
91634250005229088726469252846 6610
466539217148208013050229805263783
6426959733707053922789 153510 56888
39381132497570713310295044303467 1
5989448786684711643832805069250776
627450012200352620370946602341464
89983902525888301486781621967 7519
4583167718762757200505439 79441245
990077115205 1546199305098386 98254
28464072555409274031325716 3264079
2934183342147090412542533523 24802
19322770753555467958716383587 5018
15933871742360615511701312352563
3485820365146141870049205704372 01
8261733194715700867578539336 07862
2739558185797587258744102542 07710
547536129404746010009409544495966
28814869159038990718659 80563617 13
76922272907641977551777 2010427649
6949611056220592502420 21770426962
2154958726

45398922769766603105249808557594 71
631075870133208861463266412591148
633881220284440694169488261529577
625325019870359870674380469821942
056381255833436421949232275937221
289056420943082352544084110864545
369404969271494003319782861318186
188811118408257865928757426384450
059944229568586460481033015388911
499486935436030221810943466764000
022362550573631294626296096198760
564259963946138692330837196265954
739234624134597795748524647837980
795693198650815977675350553918991
151335252298736112779182748542008
689539658359421963331502869561192
012298889887006079992795411188269
023078913107603617634779489432032
102773359416908650071932804017163
840644987871753756781185321328408
216571107549528294974936214608215
583205687232185574065161096274874
375098092230211609982633033915469
494644491004515280925089745074896
760324090768983652940657920198315
265410658136823791984090645712468
948470209357761193139980246813405
200394781949866202624008902150166
163813538381515037735022966074627
952910384068685569070157516624 19
2987244482

7 1942933 10048 5482445454580 7 18897633
00323252582 158 1280327467962002814
762431828622 17 1054352898348208273
45 1680 186 131 7 195933247 11074662228
5087 10666 1 1770346535283957762 5997
7446721857 15816 1264 11 14327 1794347
885990892808486694914 1390977 16736
9002777585026866465405 65950394867
84 111 0790 1 16 1040085727445629 38425
494 16759460548 7 1 17235946429 105850
90995021495877931 12 19613590831 5882
6206823321 56 1530868337308381 73279
3281969838750870834838804638 84784
4 188400318 47 12697454370937329 8362
402875 19792080232 18787448828 72843
72737 8017 82700805878 24 107493575 14
889978911739746 129320351081432703
25 1409030487462262942344327571260
086642508333 18768865075642927 1605
5252 8954492 153765175 1492 196367 18 1
04943531785838345386525565664 0657
25 13635750643532365089367904 31702
597878 177 1903 1486796384082881 0209
46 14900797 15 1377 170990619549696 40
0708676671 023300486726314755 10537
23 1757 11432231741141 168062286 4206
3889062 10 192355223 5467 11662 137499
6932693217370431 0598722503945657 4
9246 16978260970253359475020913836
67377 289443869640002 81 1 034402608 4
7 128990007

46807764844088711341352503367877 3
1679770937277868216611786534423 17
3226463784769787514433209534000 16
506921305464768909850502030150448
808342618452087305309731894929 164
253229336124315143065782640702838
984098416029503092418971209716016
492656134134334222988279099217860
426798124572853458013382609958 77 1
78113102167340256562744007 2968340
661984806766158050216918337236803
990279316064204368120799003162644
491461902194582296909921227885539
487835383056468648816555622943 156
731282743908264506116289428035016
613366978240517701552196265227254
558507386405852998303791803504328
767038092521679075712040612375963
276856748450791511473131344000 18325
703449209097124358094479004624943
134550289006806487042935340374360
326258205357901183956490893543451
013429696175452495739606214902887
289327925206965353863964432253883
275224996059869747598823299162635
459733244451637553343774929289905
811757863555556269374269109471170
021654117182197505198317871371060
510637955585889055688528879890847
509157646390746936198815078146852
6213325247

3837651192990156109189777792200870
5793396463827490680698769168019749
2365624226087154176100430608904837
7976678519661891404144492527048088
197149880154205778700652159400928
9777601330756847966992955433365613
9847738060394368895887646054983887
1478968482805384701730871111776115
9663505039979343869339119789887010
9156541709133082607647406305711411
1098839388095481437828474452883836
8079418884342666222070438722887411
394780101772139228191199236540551
6395893474263953824829609036900280
83593277458550608013179884071624440
6563997948275783650195514221551330
928197822698427863839167971509126
2410548725700924070045488485692950
04481107380807996547481568913935380
0943474556972128919827177020766601
360248958146811913361412125878389
55773571949863172108443989014239400
84966592517313881716026632619310600
5366535041473070804414939169363260
237376777095850313255990095762730
195730864804246770121232702053374
266705314244820816813030639737873
664248367253983748769098060218278
578621651273856351329014890350988
32706172589325753639939790557291700
5160097615

45904477 1692265806315111028038436
01737474215247608515209901 6158582
31257159073342173657626714 2390478
27958728150509563309280266 8458937
64964977023297364131906098 2740633
5310897924642421345837 409011 69391
9642504591288134034988106 35400887
596820054408364386516617880 557608
9568967275315380819420773325 97917
27843762566118431989102500 7491829
08647514979400316070384554 9465385
9460274524474668123146879 43441610
99333890899263841184742525 7044572
51745932573898956518571657 596148 1
26602031079762825416559050 6042479
11401695790033835657486925 2800743
02562341949828646791447632 2774005
52946090394017753633565547 1931000
17543004750471914489984104 0015867
9461792416100164547165513370 74073
950260442769538553834397550 54887 1
0997852054011751697475813449 26079
43368954378322117245068734 4231989
8788441285420647428097356 25807066
98310697993526069339213568 588139 1
21480735472846322778490808 7002467
77630360555123238665629517 8853 19
67303463470122293958160679 2509153
21748903084088651606111901 1498443
41235012464692802880599613 4283511
8847154497

7127847336 1766285062 1697787177438
2436256657 1117794500644777183702219
9910669502 1656757644049979407650
3799995484500271066598781 36038023
1412683690578319046079276529 72776
940436130230517870805465 115424693
952651271010529270703066730244471
259739399505146284047674313637399
78259184541176413327906 4606365841
52927019030276017339 474 8669603486
94976541752429306040727005 0590395
0314852292139257559484507 88679779
25253931765 1564161971684435243697
944473559642606333910551268260615
9572621703669850647328126667245219
89060549880280782881429 7963366967
44124805982192146339 5657457221022
986775997467381260693670691340815
5941201611596019023775352 55563006
0624798326124988128819293 73434768
6268921923977783391073310 65882568
137771723283153290825250927330478
50724977139448333892552081 1756084
529665905539409655685417060011798
572938139982583192936791003918440
992865756059935989 10002969864 4609
74714718470101531283762631 1467742
091455740418159088000649 432378558
3930853082830547607679952 43573916
31221886057549673832243 1956506554
6085288120

190236364471270374863442172725787
950342848631294491631847534753143
504139209610879605773098720135248
407505763719925365047090858251393
686346386336804289176710760211115
982887553994012007601394703366179
371539630613986365549221374159790
511908358829009765664730073387931
467891318146510931676157582135142
486044229244530411316065270097433
008849903467540551864067734260358
340960860553374736276093565885310
976099423834738222087292464449768
456057956251676557408841032173134
562773585605235823638953203853402
484227337163912397321599544082842
166663602329654569470357718487344
203422770665383738750616921276801
576618109542009770836360436111059
240911788954033802142652394892968
643980892611463541457153519434285
072135345301831587562827573389826
889852355779929572764522939156747
756667605108788764845349363606827
805056462281359888587925994094644
604170520447004631513797543173718
775603981596264750141090665886616
218003826698996196558058720863972
117699521946678985701179833244060
181157565807428418291061519391763
0059194314

434605 1540477 105700 5433900 0182453
1177337 1895585 7603607 182860506356
479979004 13976 18089553636696 03162
19311325022385 179167 2055 180659263
51803625 12 1457592623836934822 2665
895576994660491938 112486609099798
12857 182349400 66 15552 196 112207203
0922776462009993 152442735894887 10
576623894693889446495093960330454
34084210246240 1048723328750081749
179875543879387381439894238011762
700837 1960530943839400637561 16458
560943 1295 175977 13935396074322792
48922 12670458081833 13764 1658 18269
56210587289244774003594700926 8662
6596514220506300785920024882 91860
83974373235384908396 4326 147000532
42354064704208949921025040472678 1
059083644007466380020870126664209
457181702946752278540074508552377
720890581683918446592829417018288
2330 1497 15542352359117748 18628592
96760504820386434310877956289 2925
405638946621948268711042828163893
9757 11757786915430 165058602965217
45958 19888786804081 10328432739867
198621306205559855266036405046282
152306 154594474489908839081999738
7474529698 107762014871 34000122535
522246695409315213 115337915798026
9795557 105

0850747387475075806876537644457825
2443263804614304288923593485296 10
5826938210349800040524840708 44035
611678171705 128 133788057056434506
16 1193304244407982603779511985486
945591520519600930 4127 10072778493
01555038895360338261929343797 0818
7432094991415959339636811 106275572
9527800425486306005452383915 10689
989 135788200 194 1 17865356821491 185
28207852130 1255 185 18493711 15034221
59542244511190020739353962740 0208 1
10465530207932867254740543652 7 175
95893500716336076321614725815 4076
42053020045340 183572338292661 9 153
0835409512022632916505442612361 9 1
9705 1613839357326693760 1569144299
4494374485680977569630312958871 9 1
61 12929468 188493633864739274 76012
269641588489009657 17086 160598 1472
04467428664208765334799858222 0906
198021732 1 16 14 2304 194777549907 387
385679411898246609 1309 16917722742
0723336763503267834058630 19301932
42996397204445 17928812285447 82 119
5353089891012534297552472763 57302
2628 138209 180743974867 14535907786
33530 16082 15599 1 13 14 14420509 14472
9353502223081 7 19366350934686585 86
5631485557586244781862010871 18897
6065296989

9269328178705576435143382060 14 107
73292610634315253371822433852 6352
0217735440715281898137698755 15757
45469397271504884697936195004 7772
0970561793913828989845327426 22728
86471088832701737232588182446 5843
6249580592560338 1052 15606206 15571
32991560848920643403033952622 6345
145428367869828807425142256745 180
6184149564686111163540497 189768215
42277224794740335715274368 1940989
20501136534001238467 14296551 86734
41537416150425632567 1343024765512
52192180357801692403266995417 4608
759240920700466934039651017813485
78356944407604702325407555577 6472
84507518268904 18293966 1133 10 16013
111907739863246277 82 19023650 66037
4041606724962490 13743321724645409
7412995570529142438208076098 36482
34659738866913499197840131080 1558
13439791948528304367390124820 8244
4814128095443773898320059864909 15
950532285791457688496257866588599
9179867520554558099004556461 17875
5249370124553217 1701942828846 1740
273664997847550829422802023290 122
1630102309772151569446427909 80219
0826689868834263071609207914085 19
769523555348865774342527753 119724
7430873043

619511396119080030255878387644206
085044730631299277888942729189727
169890575925244679660189707482960
949190648764693702750773866432391
919042254290235318923377293166736
086996228032557185308919284403805
071030064776847863243191000223929
785255372375566213644740096760539
439838235764606992465260089090624
105904215453927904411529580345334
500256244101006359530039598864466
169595626351878060688513723462707
997327233134693971456285542615467
650632465676620279245208581347717
608521691340946520307673391841147
504140168924121319826881568664561
485380287539331160232292555618941
042995335640095786495340935115266
454024418775949316930560448686420
862757201172319526405023099774567
647838488973464317215980626787671
838005247696884084989185086149003
432403476742686245952395890358582
135006450998178244636087317754378
859677672919526111213859194725451
400301180503437875277664402762618
941017576872680428176623860680477
885242887430259145247073950546525
135339459598789619778911041890292
943818567205070964606263541732944
6495766126

5195349570 18600 154 126239622864 138
97796733329070567376962156498 1845
068422636903678495559700260798679
96261019039331263768556968767670292
95371162528005543100786408728 9392
2571451248 1135778627664902425161 9
9027747 1090335933309304948380597 8
56628844787441469841499067 1237647
8958226329490467 98 12089984857 1635
7108783 11 9 184863025450 1620929805 8
29208334813638405421720056 12 19893
53669371336733392464416 1252231969
43471206417375491 2 1635700857369 43
973059797097 197266666422674311 177
62176403068681 31035 18991 1227 13397
2403688700099686292254646500638 52
88620393800504778276912835603372 5
482557939 1298525 150682999691077542
57647488325341412132800626717094 0
0909822352965795799780301 82824284
902214707481 111 1240 186076 134151503
87569830918652780658896682362523 9
378452726345304204 188025084423631
90383318384550522367992357752929 1
0692504326 14469501098610888999 146
58551881873582528164302520939285 2
5807796973762084563748211443398 8 1
62710031703151334402309526351929 5
8868069082 1355853680 16 1000213740 8
51 15448491268584 1268695899 174149 1
3382057849

2800698255 1957 40 20 18 18 10 564 1297 25
083607035685 10 5533 17 87 84 08 2900004
15525 1186577945396 33 17 5385 3209 214
9720526607 83 126028 1961 16485809868
4587525 12999740409 279 768 317 663 991
4655386 1089375879522 1497 173 1728 13
1517932904 43 1 12 18 1587 10235 1874075
7222100 123768721944747 2093 4931232
4107065080 6 185623725267325 4073332
48757544829675734 500 1932 1902 199 1 1
9960797989373383673242576 10393898
5349278777 47 398050808001554476406
10535222032540944356771879456543
040673589649 10 176 1077594836454082
348613025471847648518957583667439
979150851285802060782055446299172
320202822291488695939972997429747
1155371858924238493855858 59540743
81048826246487880533 0427 146301194
15898963287926783273224 56 10385219
70111304665871005000832851 7731177
648973523092666123458887310288 35 1
562644602367 199664455472760831011
8788389 151 14934093934475007302585
5814756 19088 1398752357 81233134227
9866503522725367 17 123075686104500
454897036007956982762639234410714
6584895780 24 1408 158405229 53693749
97 10665594894459246286619 96355635
0652623405 339439 14211 127 18 1069 105
2290024657

423604 13009369 18892 55865 78466846 1
21567955 42566054 16005071 27664 1766
0568742742 00329577 16064344 860620 1
23982 169827 17231978 268 16628249938
7 14995449 1373020518436690 76723577
40005393266262276032365975 17 18925
90180 1104290384274 185507 894887438
8327030632832799630072006980 12244
365 1163940869222207453 20244624 12 1
15580435454206 42 15 12 1585056896 157
3564 143130688834 43 185280853975927
734433655384 18834030 35 17822946253
702015782157373265523185763554098
954033236382 3192 19892 171 177449469
40367829618592080340 3867575834 111
5 18824177439 14507736638407 1880489
35825686854201 164503 1357633355509
440319236720 34865101056 1049872726
4721319865434354504091318595 13 145
18 12764373 104389725070049 81987052
1762724940652146 19959232 142314439
77654670835 17 147493679 86 186552791
7 1582408065106379950018 4295938799
1583501715807598837849622 57398512
1298 103263793762 1832245659 4236685
376799 113140 108043 1397323354 49090
824910499 1433258432988 21033984698
1417157560 108297065830652 11347076
803680695322 97 199059990445 1209087
27577622535 10409023928887 79424630
4832803191

327 104954 78 599 180 196 96 78 35 32 14 644
4 118 92 60 63 15 266 18 167 443 193 55 508 170
8 187 547 70 508 02 654 025 294 1092 182 648
582 138 57 52 66 88 155 584 1131 985 600 22 1
35 158 88 72 103 656 960 875 1506 318 753 30
029 421 186 822 2 1893 7755 460 272 272 912
905 04292 2597 877 1066 78 738 400 006 167
72 154 63 844 129 237 119 352 18 284 998 243
509 208 9 180 1685 572 798 1564 2 1858 191 1
9749 09857 305 703 3266 764 64 607 287 574
30565 372 60 276 898 237 325 97 450 844 796
495456480 3077 1598 15395 58 2777 91 393
73601 717 422 99 60 2735 3 1027 687 194 494
449 17939 785 1446 3 1597 31 4435 35 18 504
914 1394 15573 293 820 48 542 1235 08 1739
125497 498 193087 143 9661 513 294 20 459
19380 106 23 142 1774 199 184 060 180 3479
4988 769 105 1557 905 554 806 9538 785 400
664533 7598 1862 8464 1990 52 204 528 033
062 63 695 62 64 909 108 276 271 159 038 569
9505 124 652 99 960 6285 544 38 383 303 276
38599 8007 929 228 46 659 503 551 2 1 12 452
840 875 1622 906 026 20 1 1857 7753 137 479
49362 05549 640 107 3001 348 853 1507 354
87353 9056 029 089 335 264 007 1327 47 326
219 603 1 177 343 394 367 3385 759 1245 08 1
49335 7369 1 166454 1281 7881 714 540 230
547 506 67 1365 1825 828 489 80 995 121 39 1
93995 63324 13365 5677 70 980 0308 19 102
72040 997 148 687 4 1813 466 70060 940 510
2 146 269 028

0449 1596 4654 5330 1077 5469 54 1308 87 1
4165 3125 4481 3061 1924 0782 1188 69005
6027 7818 2423 5022 6961 8934 4352 54763
3573 5364 8561 9363 2544 1775 661 1398 170
3930 6328 72 1669 0572 2259 7452 09 1929 1
7262 1998 4440 9646 1582 6945 6380 23950
2837 1216 8644 6561 7852 3556 5164 1277 1
2826 9186 8861 5572 71620 147 493 405 227
6946 5957 12 1983 1494 338 16221 1400 693
6307 4304 44 1732 8478 610 1777 74 383 797
7037 2317 9525 5434 1072 2344 55 1255 558
9998 6461 8387 6764 9039 7246 116 795 90 1
8100 0350 9892 8641 204 195 16355 110876
3204 2676 1297 9826 5294 2588 2951 14 127
5841 2627 3279 0798 8075 5975 185 157 684
1264 7422 0947 9721 8433 0935 29726 6521
0015 6625 1455 2994 7451 2763 155 091 763
6730 2594 62 1329 30 1904 0283 7954 24632
3258 5550 3010 9670 6922 7202 2707 486 34 1
9005 4383 0265 0681 214 142 135057 15417
5057 5086 3990 7673 9463 3514 6209 08 288
8934 9383 7643 9399 2569 0060 4067 31 142
2093 3121 9593 6202 9829 7235 116 325 938
6772 2414 7791 16295 727 8075 239 505 625
1581 6031 3335 9382 311 5005 1862 689 053
0658 3681 2998 8108 6632 6327 1980 61 127
1548 8587 9809 3487 9129 1370 749 823 057
5929 0918 6293 9195 0147 21 19758 606 727
0092 5477 1802 5750 3377 3079 9397 13453
9532 6461 9526 9996 5963 8565 49 1759 045
8333 585799

10201271320458390320085387888 1633
63768518208372788513 1175227769609
787962 142372 162545214591281831798
2160441113 1167 1406914827 170981015
45778193920231156387 1950805024679
72579249760577262591332855972637 1
2112019057 2077 1409 148645074094926
7 18035815 15757 15 1405039761 0963846
7555692989703835473 1410 0223802583
4687673501297754 13279532060971154
506484212185936490997917766874774
4818828706323155 158650328981 64228
2882327468661 0659273219790 7162384
6421534898524762 1 6789050260998045
26648392954235728 7 343977680495774
091449538391 57556548545905897 6495
198513801 00795801 07837 59945775299
19670054760225255 20 34453988712538
78017 19607 18 164078 1248478472 57912
407824544361682345239570689514272
2697504318736332630 1 1103053423335
8216093331 9121880660826834 1428910
41517324721 605335584999322454873 0
77882290525232423486 153 1520976938
461 0425828497 149634753418375620 03
0 1491570327968530186863 1572488401
5266398356895636346574353217 83493
1998255421 1730846774529708583950 7
6164582296303244243282 37 737450517
028560698067889521 768 198 1567 10781
6334052667

595394249262807569683261074953233
9053622309080708 14559198373553777
487420290390 18 1429373115293346444
6815121294509759653430628421 53 194
45727118614900017650558177095 3024
68875263250119705209476159416768 7
27784472000192789137 25 18416228577
8379228443908430 118 112 14963664246
590336341945406571835447719124466
212593926566203068885200555991212
35363718226922 53 17814587925937504
4144893398160865790087616502463 51
970458288954817937566810464746 14 1
05142498870252 13993687050937 23054
47734112641 35489280684105910 77166
778212383328 1026218558775131272 11
7934444820 1440425745083063944738 3
6379390628300897330624 138061 45894
142276947479316657176231824721683
5067807648757342049155762821758 39
72975134478990696589532548940 3356
156131674032764724692 125057591162
5152965456854463349811431 76702572
95661844775487469378464233737238 9
81920662048511894378868224807279 3
520225017965453437572741639107919
729529508129429222053477173041844
7791567399 1738418311 71103625243957
16152714669005814700002633010452 6
43547865903290733205468338872078 7
3544476264

7925297690 1709 12007874 18373673508
7713376977683496344252419 94995 138
8315074877537433849458259 76556099
655595431804092017 8497 18468549737
0696212088524377013853757 5768141663
2722412634423982 15294164537800049
2507262765 15078908507 126599703670
8726692764308377229685985 169 12230
5037462744310852934305273 07886528
3977335246017463527703205938 179 12
5396915621063637625882937 57 137384
0754406468964783 1007045806 1344673
127 159 1194608435935825987 78283526
653115106504 16232953290477 72 17408
3559349723758552 138048305 09000964
6676088301540612824308740 64559443
18534137552201663058 12 11 103345312
0745086824339432 159043594 43031243
1227471385842030390 10607 094031523
5556172767994 1600203939750 9989762
9335325855575624808996 69 182986422
267750236019325797472674257821 111
973470940235745722227 121252685238
429587427350 1563660093 18804549333
8989741571490544 1825597 3808087 156
528143010267046028431681923039253
52977957658624 14392701549 74087927
313105 1636 1191375770089 2956482332
3648298263024607975875767 74537716
0102490804624301856524 16 175665560
0160859121

5345556267602 19268998 28553 77872583
1451440826545834844094784631 78777
374794653580 169960 77940 5568701 192
3286080411309046293508 71827125934
66871276669487389982459852 7786499
5691654640294589350649643 35809824
7659651651420909867552038 08309203
23048734270346828875 160407 1546653
834619611223013759451579252696743
64253192739003603860823645 0762698
82749761872357547676288995 0752114
80485252795084503395857083 8130476
93788132112367428131948795 0228066
32017002246033 198967 1970649 16374 1
17585485187848401205484467 2588851
4015627250 1982 17 1906696081 2627785
48596481836962 14 1072 17 1421498636 1
918774754509650308957099470934337
856981674465828267911940611956037
845397855839240761276344 105766751
0243075598 14552786 1678 15949657062
5597550743065210853015979 08073343
736079432866757890533483669555486
803913433720 1564988342 20893 39997 1
6414797469386969054800891 93067 138
057171505857307148815649920714086
758259602876056459782423770242469
805328056632787041926768467116266
8794634869504645074202 1937 3945259
2626686135529406247813612 06202636
4981999994

9840514386828525895634226432870 76
6329930489172340072547176418868 53
5137233266787792173834754148002 28
0339299735793615241275582956927 68
3723123479898944627433045456679 00
6203242051639628258844308543830 72
0149567210646053323853720314324 21
1260742448584509458049408182092 76
3914000854042202355626021856434 89
9414543995041098059181794888262 80
5206644108631900168856815516922 94
8620301073889718100770929059048 07
4909242714101893354281842999598 81
6966099383696164438152887721408 52
6808875748829325873580990567075 58
1701794916190611400190855374488 27
2620093668560447559655747648567 40
0817738170330738030547697360978 65
4385938218722058390234444350886 74
9986650604064587434600533182743 62
9617786251808189314436325120510 70
9469081358644051922951293245007 88
3339878842933934243512634336520 43
8581291283434529730865290978330 06
7126179813031679438553572629699 87
4035957045845223085639009891317 94
7594875212639707837594486113945 19
6028675121056163897600888009274 61
1586080020780334159145179707303 68
3519697776607637378533301202412 01
1204698860

920933908536577322239241244905153
278095095586645947763448226998607
4813297302630975028812103517723 12
44650953496536930900 18637764094 09
43498373 1325 13218620802 1480992268
55029484546618 1471 15557444709 66953
0177690434272031892770604717 78452
79391604722815343798035396798 6142
437095668322 14914654380 1459382 927
739339603275404800955223 18 1666738
0357 18393275707714204672383862461
780397629237713120958078936384 144
792980258806552212926209362393063
7313496640186619510811583471 17331
20258058667276399927635790780 6381
88130691563662741254312595899 3611
964762610140556350339952314032311
381965623632719896 1837254845333 70
206256346422395276694356837676136
8711962921818754576081617053 03159
072882870071231366630872275491866
139577373054606599743781098764980
2414011242 1427736680827513909593 1
340415582626678951084677611866595
7660165998178089414985754976284 38
78561002637965431783 1363402513581
416115 1902096499 13354873313 1 1502
270068 1930 135929595971640197 19605
36250335584799809634887 1803911 161
28135959685654788683258564378961 7
3159762002

15528962979048 19822199462269487 13
746244472909345647002853769495885
9591606789282491054410544 1251599630078
1368367490209374915732896270028 65
68293444313423473512339298259 16673
95034259958689706972673325827 3590
31212887466604514614878503461 4282
77659916080903986525757172630 8183
349444182019353338507129234577437
557934406217871133006310600332405
39916936826037461766385657588 7758
02012293663532702671006812618 2517
29146082025418928859352444910 7013
82062115538277935652969145765 0204
86432828655793470720963480737269
2141186895467322767751335690 19015
37236690368653891612916888878 7640
75254934942497334271811788927 5993
159671935475898809792452526236365
903632007085440784544797348 29 180
2082044926670634420437555325 05052
7522833778887040804033531923 40768
56301093477721256390886404 1310107
3817853338316038135280828119 04083
256440184205374679299262203769871
801806112262449090924264 198582086
17511771137890516091403815 7500336
6424156095216328 1971223350231 6742
2600567941281406217219641842 70578
43289598028823350598282081966 6624
9035857789

9403331522748177776952843681630088
53176969478369058067 1064828083598
046698841098 1351586549069333 19522
39436328792399053481098783027450O
1720654336990661177845543646877 23
631844464768069142828004551074686
64539280539940910875493916609573 1
6197150331669683099294663491427 98
78084225722069714887558063748030 8
862995118473 18712477729191007022 7
588893486939456289515802965372 150
40960310776 128983126358996489341O
247036036645058687287589051406841
238124247386385427908282733827973
326885504935874303 160274749063 129
572349742611221517417 153 1336 18622
410913869500688835898962349276317
31647834007746088665559873338211 3
8299287769 11495492 1841920877716O6
06847287467368 1886 1675072210172 6 1
10383067178785669481294878504894 3
063086169948798703 1605 15884108 282
35127415353851336589533294862949 4
49506186851477910580469603906937 2
66267038651290520 11378 1085861618 8
886947957607413585534585151768051
97333443349523012039577073962377 1
3160302428872005373209982530089 77
6189731298 178819446717 31160647231
476248457551928732782825127182446
8078242152

164695678192940982389262849437602
488522790036202193866964822156280
936053731780408637272684266964219
299468192149087017075333610947913
818040632873875938482695355830773
957614479972700034728801827852813
895032179863452161110666088393140
532269449054555278678944175792024
400214507801920998044613825478058
580484424164047750315360549065914
300781583724301231375115622840158
386442708907182848167575271238467
824595343344496220100960710513706
084618011875431207254913349942476
171156333214089346091565615506003
173842187015702261031019166038870
646614388977363187809407115275281
746895764015810470169652475577408
916445686777171585005832699434016
772021567677240681283665652641229
824394651331973591997094032759385
026695574702318132032437164205861
410336065245369391600506449530601
612678226489424373971667176612310
489750318857321655549883421218028
469125290861014855278152776256237
504563757694977343368460156077270
355096290493924870884062810679436
224187047470083688426710225583024
035998416459511224852726336326451
1401739524

8086 1946358407837535356885622317 1 1
5520947223065437092606797351000 56
554938 12245754837285457 1 1797393 6 1
5756 16764 169289580525729752233855
86 1 138832 2 17 1 10736226581621884244
317885748879810902665379342666 42 1
6990914056536432249301334867988 15
488662866505234699723557473842483
0590423677 1432787923 1642240387776
4330 192600 1922847783 1383763253612
10253369358 1262408686669973827597
73656822279072 1583247888864236934
6396 164363308730 1398 1421 143030600
8730666 1648036789840913359262 9340
230432497492688783 164360268101 130
9570716 14 191283068657732353263965
36773903 1766 136 13 1596555358499939
860056515592 19367599777 1793301974
4688 14837 1 1032065036931928 9452 140
26509 15465 18430993655349333 7 18342
52984336799 1593941746622390038952
76738 133306 1774762957494386871697
845376721949350659087571 19 1772087
5477 107 18993796089477451265475750
187 1 194870738736785890200 6 1737332
1075693302216320628432065671 19209
6950585761 1739616323262 1770894542
62146098584 102378 132 158177276 0222
2738 133495410481003073275107 79994
899 197796388353073444345753297591
4263768405

4422647842 16063 1227696 46967 156473
9990437 15903 3239065607 266441 16438
605404838847 16 19 12 10900 870 10 19 130
72607 1044 1 14 14324 1976 796828547885
5247794764 8180295973 6049439700479
5960402927 46299 20357 209976 1950 140
3483 153809 477 1460 1056333446998820
82212058728 15 10729 18297 121 19 17876
4248803546723 169 1654 18522 56729234
429 187 128 16323 25969 654 13548589577
1332083399 1 12887759 17226 115273379
0 1034 13620856 1457 799 2398 778325083
5507301998 1845902595 83559 89260553
2996737704917 2245493532 9683300002
230 18 15 17 22657 5787 5240588 32249085
8212800897 47909326 100762578770428
6560069961762 1217 6845478996440705
06624 17 102 13327 486 796 237 430 229 155
3582007801411653480 6564 7488 230615
003392068983 7947 66255036549822805
3296628621 179306284 30170492 402301
98571997894 8836897 18304 3805 182 174
419 1476604 2975 24372516834354 11217
0386313794 1142 2095 295885798060 152
9387527537 99030938 87 1683572095760
7 1522 19002 793792927 86303637268765
82268 124 1993384808 1660 21603 722 154
7 10 143007377 537 79269 90695 87 121 289
2880 1905203 160 128586 18254944 13353
82078488346531 163265040 7642428390
870 12 10 15 1

942319616522684220037112304643006
734420647477180213530701240988603
533991526679238711017062218658835
737812109351797756044256346949997
872511254408545222748109148743072
5986960204027594117894258128 18821
599523596589791811440776533543217
575952555361581280011638467203193
465072968079907939637149617743121
194020212975731251652537680173591
015573381537720019524445436200718
484756634154074423286210609976132
434875488474345396659813387174660
930205350702719529839432714253711
557666000257844230310734295515339
450604862227649666876240793243531
929926392537310768921353525723210
808898193391686682789482811704726
245019484097009757609209837240900
747179733407881418251958425980962
417476101382526439551352593118850
456362641883003385396524359974169
313228947198783084276004013680747
039040972384739458348961865397905
941185993103561684368692194853820
557803957738813606795499000851232
594425297244866667668346414021899
159445653094234406506678519484177
667794704720419588220432953803263
105374948831221803912796784461001
3972675389

2195119117836587662528083690053 24
900459741094706877291232821430463
5337283519953648274325833119 14445
90178096077828835837301 11 185754365
99589827245319253105881 1502630754
257149394302445393 1870 17992360816
66113054262539958338979429716020 7
0338767815033010280120095997 25222
228080142357109476035192554443492
99867678178910455590630 1595380976
1875920358937341978962358931 12598
3902598310267193304189215 10968915
62250696591198283234555030590 8173
07351955037216658702880539921385 7
60370353771051780212801295668 4198
41403628727256232144287543022 1090
9472721073474134975514 19073704331
82766261772759968888260272252471 3
3683353452816692779591328861 38176
634985772893690096574956228710302
43625907724122190943008717556926 2
5758065709912016659622436080 24287
0024547362036394841255954881 72727
24736534677836472019183039987 1762
703751572464992228946793232269361
91776416146187956139566995677830 6
8290316589699430767333508234990 79
06241002025061340573443006957454 7
46821756904416515406365846804636 9
2621274211075399042188716 12761778
7014258864

825775223889 18459952337629237915
585744549477361295525952226578636
462118377598473700347971408206994
145580719080213590732269233100831
759510659019121294795408603640757
358750205890208704579670007055262
505811420663907459215273309406823
649441590891009220296680523325266
198911311842016291631076894084723
564366808182168657219688268358402
785500782804043453710183651096951
782335743030504852653738073531074
185917705610397395062640355442275
156101107261779370634723804990666
922161971119425912044508464174638
589938239946517395509000859479990
136026674261494290066467115067175
422177038774507673563742154782905
911012619157555870238957001405117
822646989944917908301795475876760
168094100135837613578591356924455
647764464178667115391951357696104
864922490083446715486383054477914
330097680486878348184672733758436
892724310447406807685278625585165
092088263813233623148733336714764
520450876627614950389949504809560
460989604329123358348859990294526
400284994280878624039811814884767
301216754161106629995553668193123
2874257020

637383520200868636913117334697317
412191536332467453256308713473027
921749562270146873258678917345583
799643513588009593508775563562488
10493852999007675135513527792124124
29277488565888566513247302514710 2
105753525165118148509027504768455
182520963318990685276144351382136
62152368890578786699432288160283
774820355060160298940091197138501
798716836337441392759736440170070
1476370665570350433812111135764150
184518214136198234951596010647527
12575935185304332875537783057509 5
674254426847122196187091785607839
36144511383335649103256405733898 6
6717812397223751931643061701385 95
394743678433926709867124522111896
908402363274114966012434830989299
417380305884171666130730400675883
804321115553794406054977217059428
21514886165672771240903387727745 6
29097110134885184374118695655449 7
457368452180669829110450580042998
8795389902780438359628240942 18605
56287788428802127553884803728640 0
19441614257499904272009595204654 1
70598104989967504511936471172772 2
204361026140797508096869751766002
3718774834801612031023468056711 26
4476612374

762785219024 12025699435347 1622666
089367521983311181351114650385489
50251206557726361454736044 2685949
807439693233 1297 127377 15734 70997 1
39522911826534851555871373 3662912
02427143025037632695013509 1161295
2993785864681307226486008270 88133
35381937036825988678933212 3832705
329762585738279009782646054 559855
513183668884462826513379849 166783
940976135376625179825824966 345877
19501243840403591408492097 3375464
2474488176184070023569580 17741017
769692507781489338667255789 856458
985105689196092439884156928 069698
33522402256345704973 12245 26935419
3837004843183357 1965 16626 72157552
419340193309901831930919658292096
965624766768365964701959575473934
55143374137087615173236772 0422738
567427917069820454995309591887243
4939524094441678998846319 84550485
239366297207977452814399 41825678
945779571255242682608994086331737
1538896262889629402112 10888442737
65686245276121303710 1730078513571
54045330415079594477 76143 59743780
37424366469732471384 104921 2431413
890357909241603640631403814983148
190525172093710396402680899483257
2297954564

0427017577229041732347960736 18787
8899133183058430693948259613 18713
8164234672 187308451338772190869 75
1049428437693250249816566738 16260
6159417682525099937416728839 51744
0669325496534031014522253161 89009
2353764863784828813442098700 48096
2271712264074895719390029185 73307
460104360729190945767994614929 290
4279816877294264877299528584 34647
775386906950 1489841339245403 94 144
6802636254021186143170312511 17577
6428299146445334089209769616 99098
37265236176874560589470496817 0136
9749095230720826828878907301 90018
25342580534342170592871393 1737993
142410852647390948284596418093 614
13847583113613057610846236683 7237
695913492615824516221552134879 244
14504175684806412063652017 0386330
129532777699023118648020067 556905
682295016354931992305914246396 217
0253297475731140942201801993 68035
0264956369558664259067626856 87372
110339156793839895765565 193177883
0002416135395624377778408017 48819
373095020699900890899328088397 430
367736595524891300156633294077 907
139615464534088791510300651321 934
4866732482759079468078798194250 19
5826223203

9513125201410996053126069655540 42
4867054998678692302174698900954 78
5072567297878794769888831093487 4644
2640071818316033165551153427615 56
2240547447337804924621495213325 85
2769884733626918264917433898787 247
8927846891882805466998230368993 97
8341374758702580571634941356843 39
2939606819206177333179173820856 24
3643363535986349449689078106401 96
7407443658366707158692452118299 78
9380407713750129085864657890577 14
2683358276897855471768718442772 61
2050926648610205153564284063236 84
8180728794071712796682006072755 95
5590404023317874944734645476062 81
8954151213916291844429765106694 79
6935401686601005519607768733539 65
1161493093757096855455593815137 895
6903925101495326562814701199832 69
9220006639287537471313523642158 92
6512620407288771657835840521964 60
5410543544364216656224456504299 90
1025658692727914275293117208279 39
3775132610605288123537345106837 29
3989358087124386938593443891757 133
7630072031976081660446468393772 58
0690923729752348670291691042636 92
6209019960520412102407764819031 60
1408586355842760953708655816427 39
9534934654

6314504040 1995 28537252 00495780525

46562511154 10925243799 1326262 71360

9099402902 26206283 6752 13230506518

3934057450 11209934 14649 1843332364

656937 1725914489324 1590062420206 1

2885732926 1335968087 26500045 62828

4557574596592 1205303413 101 1182750

13069615098355 1563200431078460 190

6565493806542525229 16199 18 1995960

2752327702249855738 82489988270746

5936355768582560518068 96428537685

07720122203479209939 36 17926820659

01421656159253067379 44568 94907085

326356819683 18617722682499 1147 26 1

57320358076462981 162440133 1673789

2788689229032593349861797 02199498

19257396 176730758344 17 09855922217

01718257 1277753449 1508205 27843090

46194608352174020058386 7284970941

102326695392 14454610662 1500641067

4740207009 1899 1195 1376466 90448126

7253691537 162290 79 1385403937 56007

7835153374 1677479421 0038400230895

18509945487790393461 22220 86506016

0500351776264831611 15332558770507

35412792499098593734 73787081 19425

305512143697974991 495186053592040

3830235716352727630874 6932 1962219

0064260886 183676 10334600225547747

781364101269 1906569686 4950 1268837

6296907233

9612762872230411418136100602640 44
0300359969889199458273976241146 13
7448040596970625767647237660655 54 1
61857469052722923822827518679 9 156
98339074767114610302277660602 0006 1
24687647772881909679 16 133540 19 88 1
402757992174167678799231603963 569
492851513633647219540611171767 387
37255572852294005436178517650 2307
54469386930787349911035218253 2929
726044553210797887711449898870 91 1
51123725060423875373484125708 6064
069052058452122754533848008205 302
45045651766951857691320004281 675
805492481178051983264603244579 282
9730129105318385636821206215 53 128
866856495651261389226136706409 395
333457052698695969235035309422 454
386527867767302754040270224638 448
355323991475136344104405009233 036
127149608135549053153902100229 959
575658370538126196568314428605 795
669666221547216956208700137277 6853
696084070483332513279311223250 7 14
863020695124539500373572334680 709
465648308920980153487870563349 109
236605755405086411152144148143 463
043727327104502776866195310785 832
333485784029716092521532609255 893
2655600672124359464255065996 77 177
0388445396

18 16328796 14 4608 17789272 17 1836908
880126778207430106422524634807454
300476492885553409062185153654355
47412547615276977266776977277058
3158014121856880117050283652755 43
21480348800444297999806215790 4564
161957212784508928489806426497427
090579129069217807298769477975112
447305991406050629946894280931034
216416629935614828130998870745292
716048433630818404126469637925843
094185442216359084576146078558562
473814931427078266215185541603870
206876980461747400808324343665382
354555109449498431093494759944672
673665352517662706772194183191977
196378015702169933675083760057163
454643671776723387588643405644871
566964321041282595645349841388412
890420682047007615596916843038999
348366793542549210328113363184722
592305554383058206941675629992013
373175489122037230349072681068534
454035993561823576312837767640631
013125335212141994611869350833176
587852047112364331226765129964171
325217513553261867681942338790365
468908001827135283584888444111761
234101179918709236507184857856221
021104009776994453121795022479578
0695065329

659403839873699072407976790408267
940076187295478359634927939045769
736616434053597922192858705749574
816966940623342726197335181366260
637359825755524965098072601236682
836059283418558480269584137725589
708837899429105498003311138846034
019391661221866960584915714857335
682861495000190975911252188003964
197621635593757437180114805594422
987304181968080856472657135476128
316292004498803154021055305970766
663627493283089168809323592900817
874119857383171926167288349184024
297212904349655269427264025596414
635259143484006758676903503823205
729341329815935330444464968294413
673234421583807616948312193331198
190610961429522015361702985751055
943264614685054526849757648078080
092213358113781977492717685450755
383287688744745915937311624706010
912446098294248412875202244625944
776387494919978404468292573609685
345498432665368628444893657041118
177938064416165312236002149187687
694673984075171763075168498563592
014868929431059402024579696229245
666448819675762943495353263821716
133957577907663707645695702597388
0043841580

5894336137 10655185998760075492418
721171488929522173772114608115434
49826654798725800566747240511 2200
738345927157572771521858994694811
794064446639943237004429114074721
81802248258377360173466853007 4498
5564715420036123593397 3129 1445859
1522887408719508708632218837 28826
2822884631843717261903305777 14765
156414382230679184738603914768310
8141358275755853643597 72165002827
78037134228696887873497 9509603110
8899196143386664068450697 42078770
028050936720338723262963785603865
32164323488155575570184690890 7464
78791224363755566686780676 1054495
50172607911429308312857 6125448194
4449473244819093795369008 20638463
167822506480953181040657 025432760
438570350592281891987 806586541218
429921727372095510324225 10797 1807
783304260908679427342895573555925
272380551144043800 123904 16877 1644
518022649168164192740110645162243
110170005669112173318942340054795
968466980429801736257040673328212
99621536848814041021944634 2464622
0745575643960452985313071 40908460
84996537678037932018991408 6581466
21753193376659701143306086 2500982
9566917638

846056762972931464911493704624469
351984039534449135141193667933301
936617663652555149174982307987072
280860859626112660504289296966535
652516688885572112276802772743708
917389639772257564890533401038855
931125679991516589025016486961427
207005916056166159702451989051832
969278935550303934681219761582183
980483960562523091462638447386296
039848924386187298507775928792722
068554807210497817653286210187476
766897248841139560349480376727036
316921007350834073865261684507482
496448597428134936480372426116704
266870831925040997615319076855770
327421785010006441984124207396400
139603601583810565928413684574119
102736420274163723488214524101347
716529603128408658419787951116511
529827814620379139855006399960326
591248525308493690313130100799977
191362230866011099929142871249388
541612038020411340188887219693477
904497527454288072803509305828754
420755134816660927879353566521255
620139988249628478726214432362853
676502591450468377635282587652139
156480972141929675549384375582600
253168536356731379262475878049445
9441834291

72756988376222626 18463654527434976
624111384513054814449836311178978 44
897320767195087841586188796929558
19733250699951402601511675 5297505
754378102422389579257865621284327
31202200716730574069286869363930 1
8676595825132649914595502609170693
475194089753574640168308117988464
524736189560564794263580705625632
8118926966302647953595 1097 1276 59 1
362331808669215357886078127599 105
3717140220450618607537486630 63505
9148391646765672320571451688 61707
909846959322367249467375830996070
42589220481550799132752088583781 1
17685214269334786921895240622657 9
210436203488529262679840139532164
58791151579050460579710838983371 8
640380244175113472264725470107947
939969535546696197267632552299 146
549334996632341859514503609803440
922122067125676987234279407088570
704742931733291885238967219713539
244924261786411886377909628144869
1786946817759171715066911148002 07
5943201206196963779510322708902 95
66085562225452602610460736 13 13688
69009281721068198618553780982 0184
7115416363032626569928342415502 36
009780464171085255376127289053 350
4550613568

41437758544296779770 1466029438768
72251153638011917581540281208 1825
56064854107879335989210644272 4489
861896162941341800 1295 13068363860
9294100083 136673372153008352 69623
573717533073865333820484219030818
644918409372394403340524490955455
8016406460761581010301767488 4750 1
76619086929460987692016912021 8168
8291040870709560951470416921 14702
74133900522533408348 1287035303 102
3919699978597413908593605433 59969
707560446013424245368249609877258
13110247327985620721265724990 0346
829388687230489556225320446360263
985422525841646432427161 141981780
248259556354490721922658386366266
375083594431487763515614571074552
8016159677048442714194435 18327569
840755267792641126176525061596523
54571879566731709133 1935876162825
5920783080 18520689015150471334038
61003100559148 178521 103847545 4293
3389188444 1205 17943969970 194 11269
5119526564919594 18997541839323464
742429070271887522353439367363366
320030723274703740712398256202466
26519740901997624520561985576 2576
00087081730832883443818310700545 1
44935458854226785785519 1537229237
9555494333

4101744201696000906964156127322973

770221217951868376359082255128816
47002199234886404395915301846400
7143211863606225270115411222838023
7785389110984902013427410141215593
769965438877197485376431158229838
5331230717511329619045590079380643
27669581901484262799122179294798773
3489018684716765038273285520590823
9845298062592503521284519259279863
5935061329619467962523739725655843
157853744567558998032405492186962
88849033256085145534439166022625773
7755129162007727968526293879375303
4541810807292858919897153817973433
49618723292761474785019261145041373
27487324297058340847111233374627 4
6172746265824153242710593225062553
3023147387592517247873228814914553
915605036334575424233779160374952
50249302235148196138116256391141573
6103268449580725082734317659440543
098269765269344579863479709743124
498271933113863873159636361218623
49726140955607992062831699942007 2
054811525353393946076850019909886
553861433495781650089961649079678
1429011483876456821749140756237673
6184537751440314754112067601607263
4605568592577993220703373333989163
36950434663

9069482843662998003741452762771 65
4762382554617088318981086880 68478
537055364804693509588180253605297
4079353867651119507937328208 31462
6896007107517552061443378411 45499
5013643244632819334638905093 65457
1450690086448344018042836339 05135
7815727397333453728426337217 40657
7577107983051755572103679595 769018
89958494130195999573017901 2401939
08681356585539661941371794 4876320
79868800371607303220547423 522668
96801882123424391885984168 9722776
52194032493227314793669234 0048489
76059037958094696041754279 6137825
53781223947646147832926976 5451622
90281701100437846003875654 41517394
33960048915318817576650500 9516974
02415644771293656614253949 3688842
30517400129920556854289853 8979426
69956777027089146513736892 2061044
15481662156804219838476730 8717875
90279209175900695273456682 0265133
73111518000181434120962601 6586298
21076663523361774007837783 4237091
52644063054071807843358061 0729611
05550020415131696373046849 2133568
37265400307509829089364612 0478911
14753037049893952833457824 0828173
86441322710002968311940203 3234564
2082647327

623383029463937899837583655455991
934086623509096796113400486702712
317652666371077872511186035403755
448741869351973365662177235922939
677646325156202348757011379571209
623772343137021203100496515211197
601317641940820343734851285260291
333491512508311980285017785571072
537314913921570910513096505988599
993156086365547740355189816673353
588004821466509974143376118277772
335191074121757284159258087259131
507460602563490377726337391446137
703802131834744730111303267029691
733504770163210661622783002726928
336558401179141944780874825336071
440329625228577500980859960904093
631263562132816207145340610422411
208301000858726425211226248014264
751942618432585338675387405474349
107271004975428115946601713612259
044015899160022982780179603519408
004651353475269877760952783998436
808690898919783969353217998013913
544255271791022539701081063214304
851137829149851138196914304349750
018998068164441212327332830719282
436240673319655469267785119315277
511344646890550424811336143498460
484905125834568326644152848971397
2376040328

2126602535 166939 1408 2049 947 320 486
0216277597 9177 123475 1097 502 403078
9357599377 1509 502 175 169355 5827072
5339118923 3407 0223 832 077 585 802 137
1747783787 7839 10 15234 1320 98 489 423
4596136923 40497 998 279 304 144 463 162
7072147961 17456 9757 1968 1239 291 913
7409829258 0556 1955 207 434 243 295 982
8989805292 333664 154192 563 673 80 689
4942014712 413405 2507 220406 1794355
2525552250 0874 8790 0865 683 145 42835
1677505422 9480 3274 783 044 056 438 58 1
5919526667 58282 929 705 226 127 628 7 1 1
0401348017 87 2248 0178 9684 052 407 924
3605827424 6744 307 672 16452 70 3 1345 1
35416 7649 66890 1274 7868 01010 295 133
8626986497 482 121 1862 9040 337 691 568
5762406992 9637 2493 0972 0 1628 707 200
1898354236 9036 4149 270 2369 6 1938 547
3724803298 5504 5112 089 192 879 829 874
4678641291 5941 753 1675 602 533 435 310
6267452545 07 114 18 1483 2398 806 07 297
1402347255 207 1349 07983 98982 355 268
7239509093 65667 8789 92383 712 578 976
2487559904 4322 8895 3883 773 17 348 94 1
1227570714 1095 97900 479 1930 1046 740
7504114353 8178 2464 6307 95989 555 638
9918847737 8134 13470 70246 747 362 112
0489862269 91888 5174 562 51732 519 34 1
3520381158 6335 0 1239 1305 444 19 10073
6284475675

1416105041097350585276204448891909
78901984315485280533985777844313 9
3388399431044465669244550885946 3
1408175122033139068159659251054 68
5801313383815217641821043342978 88
261196304431113887962587460902 26 1
3090084997543039577124323061690 62
6291940392 14397402708947776637 024
8815549932245882597902063125743 69
1094639325280624 1642476868495455 3
249380176393716156368478598237 159
0238542126584061536722860713 1702 6
74740131145261063765383390315921 9
434698176053583803106128878520 515
4693363924108846763200956708971 83
67490578163085 1581381619668822 220
47570437590614338040725853862 0835
651769984267745231958241826836 982
7016023741493836349662935157685 40
61397342746470899685618170 16055 11
0488097155485911861718966802597 35
4170542398513556001872033507906 09
464212711143993196046527424050 8822
25359773481519135438571253258540 4
9394601086579379805862014336607 88
252197178090258173708709164604 527
279771535099103407364250203863 867
18220522879694458387652947951048 6
6071739022932745542678566977686 59
399234168341222746630150621553 205
026553414 6

0995249356050854921756549134830 95
890653617569381763747364418337897
422970070354520666317092960759198
962773242309025239744386101426309
868773391388251868431650102796491
149773758288891345034114886594867
021549210108432808078342808941729
800898329753694064496990312539986
391958160146899522088066228540841
486427478628197554662927881462160
717138188018084057208471586890683
691939338186427845453795671927239
797236465166759201105799566396259
853551276355876814021340982901629
687342985079247184605687482833138
125916196247615690287590107273310
329914062386460833337863825792630
239159000355760903247728133888733
917809696660146961503175422675112
599331552967421333630022296490648
093458200818106180210022766458040
027821333675857301901137175467276
305904435313131903609248909724642
792845555499134900051802957070829 1
905255678188991389962513866231938
005361134622429461024895407240485
712325662888893172211643294781619
055486805494344103409068071608802
822795968695013364381426825217047
287086301013730115523686141690837
5675747637

239763185757038109443390564564468
524183028148 1079983769 18512127201
935044041804604721626939445788377
090105974693219720558114078775989
772072009689382249303236830515862
657281114637996983 1375 17937623215
111252349734305240622105244234353
732905655163406669506165892878218
707756794176080712973781335187117
931650033155523822487730653444179
453415395202424449703410 120874072
188109388268 1675 12042299404948179
449472732894770111574139441228455
521828424922240658752689 172272780
607116754046973000803703961878779 6
694882555614674384392570115829546
661358678671897661297311267200072
971553613027503556 1678 177654 42287
442114729881614802705243806817653
573275578602505847084013208837932
816008769081300492491473682517035
382219619039014999523495387105997
351143478292339499 18793660869230 1
375596368532373806703591144243268
561512109404259582639301678017128
669239283231057658851714020211196
957064799814031505633045141564414
623163763809904402816256917576489
142569714163598439317433270237812
336938043012892626375382667795034
1693343236

0750024817574180875038847509349394
548962097404854426356371649959499
209808842947903636662975260032438
563529458447289445471662092974954
966168774141208821304770228161164
560440072363515811497297392189667
373826472047226422124201656011502
849713063327958143025160136948255
670147809357908896571349261581613
46901806965089556310121218491858058
479227206918716963163300448580201
028606578585912699746376617414639
341595695395542033146280265189511
679380745733157598460861737026878
676029436777805002446733913324316
698803540732323882818475010516413
311895370364884226902704780527424
906034920829547550540034571601840
725745369381455311753542107265578
356154998744474804273234578800618
731493415660463529797794550753593
047956872093167245365472083816858
556060438019770307642460834898761
013457093948770029461757920619525
492557571090385251714885252656710
453498134198033906415298763436954
202560802776144219143189213939088
345431317696851018401038444723489
488695209819435319065065553546173
358140455448378847525262539496658
6999205841

7652780125 34 10338 96469 81864 24300 3
4146791380 61902 80596 07854 88801 078
9705516946 21522 87730 90 10446 74624 9
7979992627 12095 16847 79568 48258 334
1402266477 21084 33624 37593 74161 053
6734041954 73896 41978 95425 33503 630
1861400951 53476 696 14762 55651 87382
3292468547 35693 58028 96011 53679 178
7303553159 37836 30822 48615 17777 054
1577576561 75935 85120 16692 94311 113
8863582159 66761 88303 26 10416 46517 1
4846979385 42262 16871 61400 12237 82 1
3779774131 26897 72667 12992 02592 20 1
7408770076 95628 34739 32201 088 15935
6286281928 56357 18933 84958 85060 385
3158179760 67947 98408 78360 97596 014
9733420572 70460 352 17906 05647 60328
5569276273 49518 22032 36 14411 25841 8
2426247712 01203 57763 88895 97431 823
2827871314 60805 35335 74494 29762 179
6789034568 16988 95535 18504 47832 56 1
6380709476 95 16990 86247 10001 97488 0
9205009521 94363 23787 19764 87033 922
3811540363 47548 86268 45956 15975 519
3765410115 01406 700 12269 27474 39388
8589943859 73024 54148 0106 12359 0803
6274585288 49356 325 15853 84383 24249
3252660875 88908 31870 07091 0023 737
7106576985 05643 39288 54337 65834 259
6750653715 00533 35144 89908 29388 773
7352051459

33304962653141514138612443793 5885
070944688045486975358170212908490
787347806814366323322819415827345
67135644317153796781805819 5852464
840084032909981943781718177302317
003989733050495387356116261023999
433259780126893432605584710278764
901070923443884634011735556865903
58524491937018104162620850429925
86974358170981338940459344719374 9
3877624232409852832762266604942 38
5129709453245586252103600829 28664
972417491914198896612955807677097
95947953060131191590117739431042 0
904907942444886851308684449370590
902600612064942574471035354765785
924270813041061854621988183009063
4588187038755856274911 58737542106
4667951346487586771543838018 52134
82819158124625993351601989355 916
796893285220582479942103451271 587
716334522299541883968044883552975
3361286837225935390079201666 94133
90911687588039888288692160023 7325
7361588207163516271332810518 18760
21048521806755266486739089009071 9
513805862673512431221569163790227
7328705410842037841525683288 71804
6987952513073266340278519059417 33
892035854039567703561132935448258
5628287610

6106982297214209619935093313121717
18789107876687204454887608941017 4
798647137882462153955933333275562
0094395804345379197822805903959 59
927436913793778664940964048777784 1
748336432684026282932406260081908
0818043909145563519368560630450 89
1422896452199877988493474777291 32
7972660276584016678901364905087 41
1421268619698620441269652829 81087
0454798615595453380212011556 46979
976785738920186243599326777 689454
0605082188382279098336271671 24490
0267611784982643770330020818 44590
0097172352043319947082420987 71514
4497510170556430295428218196 70009
2025156158441742059336581481 34902
6931115170938722600264586305 61325
605792560092733226557934628 0805683
44392137368840565043430739657 406 1
0177793701414246154930707413 60805
4421002956000956635889778992 67630
517718781943706761498217564 186590
116160865408635391513039201316805
76903417259645369235080641744 6562
35152392905040947995318407486215 1
210561833854566176652606393713658
80252166622357613220194170 1372664
96607325201077194793126528276 3302
413805164907174565964853748354669
1945235803

1530 1969 1604 8099 4606 8149 040 378 198
2973 2360 9300 87 1357 6079 862 14 254 220
9641 9004 3679 0547 9049 9300 783 724 215
8195 4535 4183 7112 9368 6584 305 538 427
1762 8035 2791 2882 1129 3083 515 756 565
9994 4741 7884 3838 1565 1484 342 298 587
0424 5592 4346 9329 5232 82 1803 5083 337
2628 3791 8302 1659 1836 18 1554 217 1574
4846 5778 4201 3432 9982 5945 668 845 582
6617 1979 0121 8084 9480 3324 487 872 581
8377 4805 5222 68 15 10 1137 17 453 684 178
7028 0274 4524 4290 5474 5182 346 749 195
6418 8551 2444 2133 7783 52 142 386 59 799
2598 8203 2870 85 1093 3838 682 990 65 719
9461 4906 2902 5742 7686 0388 505 110 326
3854 4540 4191 8495 8866 5385 450 405 713
2362 9681 0691 4681 4847 8696 59 1668 618
4275 6798 4600 4186 8762 2980 555 629 630
4595 3227 9230 516 1672 159 1968 6758 495
2363 5298 9357 8850 774 608 1537 32 14 546
4298 4792 3105 1167 6357 7494 946 229 525
6949 7660 3594 7396 2430 9953 433 104 049
9420 9677 8838 2700 27 1447 8494 069 0370
7324 9106 4441 5169 6053 2565 605 867 787
5741 7472 1108 2743 5774 3151 940 607 579
8356 3629 1433 2639 78 1221 894 628 74 477
9811 9807 2256 467 1466 4054 850 13 100 96
5678 6314 8800 9030 3749 3388 753 641 831
6513 4982 5466 9467 33 16 11 81 233 648 543
9764 9325 0261 7954 9357 2043 054 021 829
7487 125 110

740401161140589991109306249231281
311634054926257135672181862893278
613883371802853505650359195274140
086951092616754147679266803210923
746708721360627833292238641361959
412133927803611827632410600474097
111104814000362334271451448333464
167546635469973149475664342365949
349684588455152415075637660508663
282742479413606287604129064491382
851945640264315322585862404314183
866959063324506300039221319264762
596269151090445769530144405461803
785750303668621246227863975274666
787012100339298487337501447560032
210062235802934377495503203701273
846816306102657030087227546296679
688089058712767636106622572235222
973920644309352432722810085997309
513252863060110549791564479184500
461804676240892892568091293059296
064235702106152464620502324896659
398732493396737695202399176089847
457184353193664652912584806448019
652016283879518949933675924148562
613699594530728725453246329152911
012876377060557060953137752775186
792329213495524513308986796916512
907384130216757323863757582008036
357572800275449032795307990079944
2541108725

69318801466793559583467643286887 6
966610097395749967836593397846346
959948950610490383647409504695226
063858046758073069912290474089879
166872117147527644711604401952718
169508289733537148530928937046384
420893299771125856840846608339934
045689026787516008775461267988015
465856522061210953490796707365539
702576199431376639960606061106406
959330828171876426043573425361756
943784848495250108266488395159700
490598380812105221111091943323951
136051446459834210799058082093716
464523127704023160072138543723461
267260997870385657091998507595634
613248460188409850194287687902268
734556500519121546544063829253851
276317663922050938345204300773017
029940362615434001322763910912988
327863920412300445551684054889809
080779174636092439334912641164240
093880746356607262336695842764583
698268734815881961058571835767462
009650526065929263548291499045768
307210893245857073701660717398194
485028842603963660746031184786225
831056580870870305567595861341700
745402965687634774176431051751036
732869245558582082372038601781739
405175130 4

3799486882232004437804310317092 10
3426167499800007301609481458637 44
8877852227307633049538394434538 27
7060876076354209844500830624763 02
5357278103278346176697054428715 53
1534001649707665719598504174819 90
8720149087568603778359199471934 33
5277294728553792578768483230110 18
5936580071729118696761765505377 50
3029303383070644891281141202550 61
5089641100762382457448865518258 10
5814034532012475472326908754750 70
7857765973254284445935530449920 700
1453874894822655644222369636554 4 1
9422544133821222547749753549462 48
2768053333698328415613869236344 33
5855386847111143049824839899180 3 1
6545863828935379913053522283343 0 1
3795337295401625762322808113849 94
9187614414132293376710656349252 88
1452823950620902235787668465011 66
6009738275366040544694165342223 90
5210831458584703552935221992827 27
6057482126606529138553034554974 45
5147034493948686342945965843102 4 1
9078592368022456076393678416627 05
1855517870290407355730462063969 24
5330779578224594971042018804300 0 1
8388142900817303945050734278701 3 1
2446686009277858181104091151172 93
7487362788

78749074652855654347488868310641 1
0051023020875107768918781525622 73
525155037953244485778727761700196
48537035551676552091193393437628 6
628461984402629525218367852236747
5108809781507098978413086245881 52
266096355140187449583692691779904
712072649490573726428600521140358
123107600669951853612486274675637
589622529911649606687650826173417
84847893372950567390078786 1792535
1440621045366250640463728815698 23
231750059626108092195521 115085930
295565496753886261297233991462835
847604862762702730973920200143224
870758233735491524608560821032888
297418390647886992327369136004883
7436615223517058437705545210815 51
33612621429118156153017588825735 9
489250710887926212864139244330938
3797333867806131795237315266773 82
08580247014335270092438032669517 4
21195076708843263464424749 12755890
7746863582162166042741315170212 45
8586056233631493164646913946562 49
747174195835421860774871 1105733845
843368993964591374060338215935224
359475162623918868530782282176398
32373061802042465604775279431047 9
61897242995330297924974816840528 9
3791044947

00459086499 18727273454 13508 101983
88186467360939257 19305 11968645601
8557824502 1823 1065889437986522432
050677379966 19695547 2440585922417
953006820451795370043472451762893
5667705084902131077366 2575 1697335
52746230294303 120359626095342 3574
3972496592 11010657817826 108 745318
874803187430823573699195 156340957
1627009924449 29749 10548985 15 19658
664740148225 1063353679497 37142510
2293418825851173719944991 15097583
746130105505064 19772 1531929354875
37119163026203032858865852848 0193
509225875775597425276584011721342
323648084027 1433563675420463 75182
55252494432965704386 1387865901965
7388028684018940876728 1671 4137033
6617326501205786539157807030 88714
261519075001492576 1129276751 93096
728453971160213606303090542243966
320674323582797889332324405779199
278484633339777737655901870574806
828678347965624146 10289950848 7399
692970750432753029972872297327934
4429886464 12725348 1606037797 07298
299173029296308695801996312413304
9393504933254 1235507 1054461182591
141116454534710329881047844067780
13807713146540009938630648 1266614
3308582068

113958383 19 1695455582594 268957698
41428893743434670841079463 18932539 1
0696395578070602 12459748982935646
135607889834724 199794785643620420
946134123876 13 19886535235831 29968
6226894860840845665560687695450 12
744866314050547353517468730098063
2278046891224682 146080672762770 84
0240226615548502400895289 16571 176
174390203375848778429 11 2896232470
59 19 187469 10420058483261406773337
5 1027 19565399469 7 1625 172483 122306
3391 93287079838007484857265 16 1234
349332733566644733585564302352808
8392434827876088616494328939 91663
992 10488307847777048045728491 4563
03353265070029588906265915498509 4
07972767567 1297950 1009822947 62289
6 189 159 144 1520032283878773485 1309
7908 10191292672271037788980539641
56362364169 1549857684083984688616
84375407065 12 103906250612810766 37
99047908879674778069738473 1704752
53442 15639038720 12388063236880370
17949308954900776331523063548374 2
568 1665336 160664 198003018828712 37
67481898330246836371488309259 2833
759022789425880600872860388591688
497306939480205 1 122 1766359 1382515
24278670094406942355 1202015683777
7885182467

002565170850924962374772681369428
435006293881442998790530105621737
545918267997321773502936892806521
002539626880749809264345801165571
588670044350397650532347828732736
884086354000274067678382196352222
653929093980736739136408289872201
777674716811819585613372158311905
468293608323697611345028175783020
293484598292500089568263027126329
586629214765314223335179309338795
135709534637718368409244442209631
933129562030557551734006797374061
416210792363342380564685009203716
715264255637185388957141641977238
742261059666739699717316816941543
509528319355641770566862221521799
115135563970714331289365755384464
832620120642433801695586269856102
246064606933079384785881436740700
059976970364901927332882613532936
311240365069865216063898725026723
808740339674439783025829689425689
674186433613497947524552629142652
284241924308338810358005378702399
954217211368655027534136221169314
069466951318692810257479598560514
500502171591331775160995786555198
188619321128211070944228724044248
115340605589595835581523201218460
5820563592

69930347885113206862662758877 1446
035996656108430725696500563064489
18759946659677284717 15395736 12 108
180841547273 14266 1748933 134 174632
662354222072600 1460 127 0 1206934639
52056444554329 1662986660783089068
1187900908 15295063626782075614388
8 1578 135 1 1346953663038784 12092346
9428687308393204323338727754 96805
2 1030282 15443247233888452 15343727
25012858974769 14608083 14404 125868
18 1540049 18777228786980 1853454537
006526655649 1709 15429522756709222
2 17474 1 1206272065662298980603289 1
67206874365494824 6 1086973672255 47
404812889242 47 18543236057534 11672
850757552057 13 1 1566979545848 87398
742228 135887985840783 135060548 29
055 1482785294489 1 12 1905383 19562422
87 1948475940785939804790 1094 19 407
067 176443903273071213588738504999
363883820550 1683402 7749607027684
488028 1912220636888636811 04356952
930065219552826 152699127 163727738
84 1899328713056346468 822739828876
3 198645709836308917786487708667618
54856800476725526754 14742851028 14
580740 3 1529921978 14557775684368 111
0 185 3 17498 1670 1642664788409026268
28244482580275320945499 15 10451851
77 16546311

80490456798571325752811791365 6278
15811128881656228587603087597 4963
84943527567661216895926148503 0785
36204527450775295063101248034 1804
58405943292607985443562009370 8091
82152392037179067812199228049 6069
73823874331262673030679594396 0954
95718957721791559730058869364 6845
57667609245090608820221223571 9254
53671519183487258742391941089 0444
11595993276004450655620646116 4655
66548759424736925233695599303 0355
09581762617623184956190649483 9673
00203776387436934399982943020 9147
07361894793269276244518656023 9559
05370512897816345542332011497 5994
89627842432748378803270141867 6952
62118097500640514975588965029 3004
86760520801049153788541390942 4531
69171998762894127722112946456 8294
86028149318156024967788794981 3777
21622935943781100444806079767 2429
27624951078415344642915084276 4520
00204276947069804177583220909 7020
29165734725158290463091035903 7842
97757265172087724474095226716 6306
00546971638794317119687348468 8738
18665675127929857501636341131 4627
53049901913564682380432997069 5770
15078933772865803571279091376 7420
8056554936

www.ingramcontent.com/pod-product-compliance
Lightning Source LLC
Chambersburg PA
CBHW022047190326
41520CB00008B/724